Developing Fieldwork Skills

A Guide for Human Services, Counseling, and Social Work Students

Brooks/Cole Titles of Related Interest

The Counselor Intern's Handbook
Christopher Faiver, Sheri Eisengart, and Ronald Colonna (1995)

Supervisory Relationships: Exploring the Human Element
Tamara Kaiser (1997)

First Steps in the Art of Intervention: A Guidebook for Trainees in the Helping Professions
David G. Martin and Allan D. Moore (1995)

Invitational Counseling: A Self-Concept Approach to Professional Practice
William W. Purkey and John J. Schmidt (1996)

Process Variables: Four Common Elements of Counseling and Psychotherapy
Frederick S. Walborn (1996)

Developing Fieldwork Skills

A Guide for Human Services, Counseling, and Social Work Students

Rosemary Chiaferi
California State University, Fullerton

Michael Griffin
Private Practice

Brooks/Cole Publishing Company
I(T)P® An International Thomson Publishing Company

Pacific Grove • Albany • Belmont • Bonn • Boston • Cincinnati • Detroit • Johannesburg • London Madrid • Melbourne • Mexico City • New York • Paris • Singapore • Tokyo • Toronto • Washington

Sponsoring Editor: *Eileen Murphy*
Editorial Assistant: *Lisa Blanton*
Production Editor: *Marjorie Sanders*
Production Service: *Scratchgravel Publishing Services*
Marketing Team: *Christine Davis, Jean Thompson*
Manuscript Editor: *Jane Townsend*
Interior Design: *Vernon T. Boes*
Cover Design: *Sharon Kinghan*
Typesetting: *Scratchgravel Publishing Services*
Printing and Binding: *Edwards Brothers, Inc.*

For more information, contact:

BROOKS/COLE PUBLISHING COMPANY
511 Forest Lodge Road
Pacific Grove, CA 93950
USA

International Thomson Publishing Europe
Berkshire House 168-173
High Holborn
London WC1V 7AA
England

Thomas Nelson Australia
102 Dodds Street
South Melbourne, 3205
Victoria, Australia

Nelson Canada
1120 Birchmount Road
Scarborough, Ontario
Canada M1K 5G4

International Thomson Editores
Seneca 53, Col. Polanco
México, D. F., México
C. P. 11560

International Thomson Publishing GmbH
Königswinterer Strasse 418
53227 Bonn
Germany

International Thomson Publishing Asia
221 Henderson Road
#05-10 Henderson Building
Singapore 0315

International Thomson Publishing Japan
Hirakawacho Kyowa Building, 3F
2-2-1 Hirakawacho
Chiyoda-ku, Tokyo 102
Japan

Printed in the United States of America

10 9 8 7 6 5 4 3

Library of Congress Cataloging-in-Publication Data
Chiaferi, Rosemary.
Developing fieldwork skills : a guide for human services, counseling, and social work students / Rosemary Chiaferi, Michael Griffin.
p. cm.
Includes bibliographical references and index.
ISBN 0-534-34654-5 (alk. paper)
1. Social service—United States—Field work. 2. Social work education—United States. 3. Social workers—In-service training--United States. I. Griffin, Michael, 1953– . II. Title.
HV11.C493 1997
361'.007223—dc20 96-33371
CIP

In recognition of the profound contributions
made by dedicated human services, counseling,
and social work fieldwork students.
We hope this book will be a source of
guidance and inspiration to you.

About the Authors

Rosemary Chiaferi, M.A., M.F.F.C.

Rosemary Chiaferi draws on an extensive background of training and practice in diverse human services settings, including community mental health clinics, hospital-based chemical dependency programs, and the Orange County Department of Children's Services. As the former director of an intensive therapy program at the Child Guidance Center of Orange County, Ms. Chiaferi provided direct psychotherapy to families, children, and groups in addition to managing program administration and design. A licensed marriage and family therapist with a Master of Arts degree in Counseling Psychology, she has been a clinical supervisor to undergraduate human services students, pre-doctoral psychology interns, and graduate students pursuing degrees in marriage and family therapy and counseling psychology. For the past six years, Ms. Chiaferi has been a lecturer on the faculty of the Department of Human Services and the graduate Department of Counseling at California State University, Fullerton. Ms. Chiaferi has been in private practice for ten years as co-director of the Center for Family and Child Counseling in Laguna Niguel, California, and is currently studying at the Southern California Psychoanalytic Institute in Los Angeles.

Michael Griffin, M.S.W., L.C.S.W.

An experienced clinician with children and adults, Michael Griffin has a background in direct practice and has served as an administrator and clinical supervisor in various mental health settings. These include clinics affiliated with the Department of Mental Health Children's Services in Orange County and Children's Hospital and Health Center of San Diego. Mr. Griffin has supervised social work and psychology fieldwork students from a number of academic institutions, including San Diego State University and the University of Southern California. A diplomate in clinical social work, Mr. Griffin is a graduate of the U.S.C. School of Social Work. He is currently the co-director of the Center for Family and Child Counseling in Laguna Niguel, California. In addition to his present clinical practice, Mr. Griffin is pursuing a juris doctorate at Chapman University School of Law.

Preface

Fieldwork and internship experiences are among the most critical components of professional development for those in the helping professions. They are times of discovery and times for asking questions: Did I make a wise career choice? Will I find that my personal characteristics are well suited to the profession? Have I correctly estimated my abilities or will I encounter many situations that cause me to reevaluate my skills? Will I find that I am more suited to interact with a given client population than I had thought myself to be? How gratifying can I expect my initial practical experiences to be? How competent will I feel as a beginner and how will I respond to critical feedback from others? Will I be able to manage crisis situations? Can I tolerate individual differences and resolve conflicts effectively? Have I acquired sufficient knowledge to intervene in a helpful manner? Do my interpersonal skills convey to others that I am there to help? The list of concerns goes on and on.

Most students anticipate their fieldwork and internship placements with both eagerness and trepidation. Having spent so much time in the academic environment, they have a powerful appetite for opportunities to apply theoretical knowledge and translate intellectual discussion into direct action. At the same time, they feel a great deal of apprehension about taking on new responsibilities and testing their skills. However well prepared they may be from a classroom standpoint, few have the advantage of being equally prepared to enter the "real" world of human services.

Through the years, students at the undergraduate and graduate levels have expressed to us their frustration with entering into the fieldwork portion of their curriculum with little or no orientation to the types of dilemmas they undoubtedly will face. We have heard the same complaint from students who have trained in diverse human services settings. It is common for them to experience the confusion and stress of new beginnings and unfamiliar surroundings without the benefit of sufficient advance preparation. Without this foundation, students often suffer discouragement and a loss of self-confidence in the early weeks and months of their placement. Students report that their initial efforts are hampered by circumstances and developments that seem beyond their control, and this sometimes results in decreasing enthusiasm and interest. What was envisioned as a potentially exciting and stimulating experience may come to be seen as merely a certain number of hours that must be accumulated within a designated period of time. An appetite for learning and exposure to challenging tasks may be gradually eroded and replaced by a vague hope that the next experience will be better.

Having witnessed this predicament numerous times over the course of several years as on-site fieldwork supervisors, we became motivated to provide a guide for students that could ease the transition from classroom to practical work settings. In this text we describe and discuss multiple strategies that can be implemented to increase the potential for satisfying early experiences in the human services professions. We draw upon our own years as students, interns, and supervisors to construct a blueprint of predictable dilemmas that seem to arise repeatedly in a variety of settings. Many of these involve interpersonal interactions as relationships with supervisors, staff, and clients are negotiated in the early days of placement in a new situation. Others focus on legal and ethical issues that are woven into the matrix of dealing with people in complex circumstances and diverse social conditions. In addition, we present a number of predicaments that often develop in the context of coping with matters of great consequence in our society, such as homelessness, poverty, AIDS, cultural differences, addictions, and death.

Because this book is designed for students entering fieldwork and practicum settings, it takes an applied focus. Through the use of descriptive vignettes, a wide variety of commonly occurring dilemmas and problem-solving strategies are presented. Each vignette centers on a particular intern's experience, and we hope that students will find it easy to relate to the events that are depicted. All anecdotes are followed by a set of questions to stimulate thoughtful reflection and creative responses and by a brief critical analysis.

It is not our intention to suggest that any singular strategy is the correct approach to the situations introduced in this book. To the contrary, the human services professions are characterized by a vast range of intervention styles that are influenced by individual traits and unique perceptions. Diligent attention to the complexities of dealing with human beings requires the ability to adapt to constant change while maintaining a stable foundation of effective problem-solving strategies. This book focuses on helping interns to anticipate typical problem areas and develop a framework for approaching these difficulties with greater confidence. We encourage the reader to use the questions following each vignette as points of departure for creative reflection and group discussion. The critical analysis offers a commentary on the significant issues and reviews potentially successful responses. These critiques explore a variety of options open to interns facing challenging predicaments, with an emphasis on the immediate and long-term results of chosen courses of action. They are designed to stimulate recognition of interpersonal styles that may be familiar but often fail to yield satisfactory outcomes. It is our hope that this process of thoughtful examination will lead students toward increasingly productive and gratifying internship experiences.

Acknowledgments

In the course of writing this book, there have been several individuals who have offered their guidance and expertise to the authors. First, our thanks to the reviewers from a variety of academic institutions, who supplied initial general commentary and specific suggestions that helped us to refine our ideas about content to be included in the text: Samuel T. Gladding, Wake Forest University; Cindy Juntunen, University of North Dakota; Mary Lou O'Phelan, Century Community & Technical College; Rich Reiner, Rogue Community College; Myrna Thompson, Southside Virginia Community College; and Mary E. Zwaanstra, Calvin College.

Along the way, there have been a number of professionals in education, social work, counseling, nonprofit agency administration, human services, and mental health who have contributed their time and valuable feedback to this project. Their input served to enhance our efforts to maintain a focus throughout the book that would make it truly useful to fieldwork students and fieldwork supervisors. In addition, several practicing clinicians graciously offered their time to review various chapters; we wish to acknowledge their considerable support. We cannot fail to recognize the students of the Human Services Department at California State University, Fullerton, for their constructive criticism as the first users of these materials in fieldwork seminar settings. Their thoughts, feelings, and actual practicum experiences contributed greatly to the shaping of the many vignettes used to illustrate typical dilemmas that students are likely to confront in a variety of fieldwork environments.

We wish to express our sincere appreciation to Gerald Corey of California State University, Fullerton, for encouraging us to undertake the project and providing ongoing support at each step. To Patrick Callanan, also of California State University, Fullerton, our thanks for investing time and energy to review chapters in the first few months of text development. The knowledge they shared with us is drawn from their own successful careers as writers, educators, and psychotherapists and has been invaluable to us.

Finally, we would like to express our gratitude to all those at Brooks/Cole whose guidance and support helped us turn our ideas into a finished product to be proud of. We thank Claire Verduin for the assistance she offered to our project from the very start. We wish to acknowledge Eileen Murphy for her generous availability throughout the working stages and the review process. Our thanks also to the staff of Scratchgravel for working so diligently on the editing and production of this manuscript.

Rosemary Chiaferi
Michael Griffin

Contents

Chapter One

Beginning Your Fieldwork Experience

- Dynamics of Beginnings
- Coping with Pressures
- Advance Planning and Preparation
- Vignettes
- Suggested Readings

Dynamics of Beginnings

Whenever we embark on an important journey or head in a new direction, we experience a range of thoughts and feelings about that new beginning. Most of us wonder where the journey will lead and whether we will arrive at our intended destination. We also may begin to consider whether we will succeed in meeting the challenges and accomplishing the tasks we will face along the way. We are reminded of similar experiences that represent success or failure. Most of us hope to demonstrate our competence by meeting or surpassing the expectations of both ourselves and others. In moments of reflection we may imagine ourselves exceeding all expectations and feeling exhilarated at the thought of all we have accomplished. We also may worry about failing or suffering embarrassment and humiliation. Beginnings usually provoke uncertainty and anxiety, anticipation and excitement. Our reactions have to do with external reality and with our own internal expectations.

The beginning of your fieldwork experience is not unlike the start of any important challenge in your life. As a professional in training, you have invested a considerable number of years preparing for this moment, and you have hopes of continuing on in a successful fashion. To a significant degree, the fieldwork experiences you will have during the coming year will be an outgrowth of the manner in which you begin the process.

Coping with Pressures

Students who enter fieldwork while enrolled in a graduate or undergraduate program find themselves in a demanding situation. To begin, you are in an unfamiliar environment and are unacquainted with the staff and with the other students. You probably are keenly aware that this will be one experience that determines whether several years of university work and, quite likely, many thousands of dollars of investment, have been a huge mistake. In just a few days you will be faced with meetings, interviews, clients, patients, supervisors, professional staff, and competing students, all in the context of physical surroundings that are totally foreign to you. Welcome to fieldwork! As you strive to maintain a measure of calm, remember that everyone requires some time to accommodate and adapt to unfamiliar surroundings. Most likely, you will need to integrate a vast amount of information over a short period of time. In some instances, you may be amazed at your ability to synthesize sights, sounds, names, places, and numerous bits of disjointed information. On other occasions you may feel overwhelmed, irritated, or lost.

In cirumstances where we feel a lot of pressure, all our usual physical and emotional reactions to stress become especially evident. For example, if you develop physical symptoms when under pressure, be aware that your internship very likely will generate the same symptoms that you've seen before. Therefore, one strategy is to become aware of your own responses early on and begin to plan ways of taking care of your needs. How you best approach that is a personal matter; there are a number of stress management techniques that you can try. A moment of reflection will probably reveal one or more ways that you've learned to find relief in the past. The point here is to acknowledge that the anxiety that you experience in your initial days as an intern is a reasonable response to a very stressful set of circumstances. Recognition of the stressors is important in order to maintain perspective, allow for realistic appraisal of your performance, and keep your sense of humor. It will be useful to consider how you ordinarily approach beginnings in your life. Do you tend to become obsessive and plan for every imaginable contingency in order to limit the opportunity for failure? Or are you more likely to dive into a new situation with little advance planning, hoping that good luck will carry you through any unforeseen difficulty? Neither approach is likely to be effective for most students.

It is important to remember that you have the ability to affect your experience. Although you may not be able to control certain factors, such as the physical location of your internship site or who is assigned as your immediate supervisor, you can determine how you react. In order to be most effective, you will need to plan in advance, become increasingly aware of your personal style and learn to monitor your reactions and perceptions throughout the internship. These skills, which will develop over time, will prove to be of value later, in other settings. If you already are practiced and competent in these areas, then you are fortunate and ahead of the game. However, few professionals fail to benefit from continued work on self-awareness, effective planning, goal setting, self-appraisal, professional interactive style, and so on. (In fact, most of us are painfully aware of colleagues who need considerable help in these areas.) Ask yourself what you are realistically capable of and set your goals accordingly. If you are someone who generally needs a little extra time to integrate complex information, then allow for the fact that you probably will not respond differently in this circumstance. If prior experience tells you that extra rest, sleep, or social support improves your performance in times of stress, pay attention and make sure you provide those things for yourself now.

Develop Reasonable Expectations

It often is difficult for interns to know how quickly they should expect to master a new set of complex skills. It generally is not reasonable to expect that you will master all of the skills important to your fieldwork in a brief period of time. Keep in mind that the supervisors you consider to be your role models are for the most part experienced practitioners who have the benefit of many years of practice. Any administrator remembers the painful mistakes, errors in judgment, and outright failures that he or she experienced as part of the slow and arduous process of developing into a competent manager. Seasoned therapists will recount examples of their own bumbling, fumbling, and uncertainty as they gradually developed expertise in the craft of psychotherapy. Mistakes and uncertainty are expected of all trainees. This is not to suggest that you should adopt a casual or cavalier attitude regarding your performance; supervisors and academic overseers notice and frown on a lack of genuine effort. However, an honest error made by a trainee who is hardworking, serious, and invested in the internship will not be condemned. The bottom line is that you as an intern will not be expected to be a master practitioner of your chosen profession. You will be expected to strive to become progressively skilled over time, with practice and the benefit of guidance. You will need to allow yourself to be in the apprenticeship of another or possibly several supervisors. Because students with a variety of backgrounds enter into fieldwork, their experiences as beginners are never identical. If you have enjoyed previous success in another career or professional setting, entering the fieldwork setting as a "beginner" may be something of a shock. If you are accustomed to giving directives and assignments at work rather than receiving them, be prepared to deal with some negative feelings when this changes. Consider that your supervisor or others who will be assigned to oversee your work may be older or younger than you. Students sometimes are surprised to find themselves under the supervision of a much younger person, yet this is a strong possibility for a student who is returning to school after a period of absence. How will you react? Whether you are prepared or not, you will need to heighten your ability to be cognizant of your thoughts, feelings, and reactions in order to effectively cope. Those of you who already possess a strong degree of self-awareness are at an advantage as you begin your professional journey. Those of you who need more time and practice: Give yourself ample opportunity to learn and grow.

Increase Your Self-Awareness

The degree to which you are able to project a competent, composed manner to clients, staff, and other professionals will depend in part on your capacity to perceive and manage your own anxiety. Without this ability, you cannot possibly manage your reactions. Being able to recognize your own anxiety is one aspect of becoming more self-aware.

As we've said before, internship is at times extremely demanding and stressful. Some of that stress is a reflection of the kind of work you are doing. Other times stress is a consequence of how you think about your performance. All of us continually reflect on our performance; this thought process is known as *internal dialogue*. The more expert you become at identifying your internal dialogue, the more your self-awareness will increase, as well as your capacity for self-control. Give yourself permission to be nervous or to feel insecure. If you notice your mood spiraling out of control, however—if symptoms develop or if you begin to dread going to your fieldwork setting—it's time to ask for help. Ask yourself what specifically is worrying or bothering you, and then analyze whether the event that you fear (such as failing your internship, or whatever) is a realistic possibility. Don't be afraid to ask other students their opinions, and remember that faculty advisors are quite familiar with the problems that often overwhelm students. Don't feel you have to handle your problem all on your own.

Advance Planning and Preparation

Learning about the Setting

If you are fortunate enough to know your fieldwork setting in advance, it is well worth the time and effort to visit the location. Being able to get a firsthand look at the offices and working areas can help immensely in preparing for the experience. We all generate stress and worry when we attempt to anticipate the unknown; gathering some preliminary information can be very reassuring. If you can meet some of the critical players in advance, such as your supervisor, professional staff, and other students with whom you will be working, by all means do so. Beginning your first week with some of the initial introductions already accomplished can make your first day that much easier. If, however, you are in the position of most students, you'll have to cope with an onslaught of information and introductions without the advantage of a warm-up. Don't worry, that's the usual route for most of us. Just try to think of each new challenge as an intriguing opportunity to learn and demonstrate your creative ability.

As an experienced student, you already are an expert at coping, adapting, assimilating complex information, and fashioning creative solutions. These are inherent strengths that you bring to the setting and that serve as the foundation for your growth. Take a moment and consider all of your personal resources. What talents, skills, and aptitudes do you have that help to define you as a competent, valuable professional? Although you may be tempted to think of yourself mainly as an insecure student, you probably already have a great deal to offer. If you cannot now fill an entire page of notebook paper with a description of your specific skills and abilities, perhaps you are not fully assessing your capabilities. Remember, you would not have been selected for your professional program had you not demonstrated significant accomplishments *and* potential. The facility that accepted you as an intern stands to benefit by your presence in a multiplicity of ways. Hospitals and other facilities that participate in student training may receive additional funding or staff as a consequence, and often are viewed as more prestigious if they have an established training program.

Supervisors have many reasons to value their students. The time spent teaching and training often provides a welcome respite from the unrelenting pace of daily meetings, clients, and scheduled activity. Most of the time, supervisors have asked to be involved in training and have spent significant time preparing themselves for the task. In general, they look forward to their work with students and value the opportunity to participate. Successful supervisor-student relationships often lead to enduring professional associations and may generate future employment opportunities.

Any new employee or intern has to become acclimated to an organization that is to some degree foreign. There are rules and regulations to learn, operational policies to digest, and hierarchical matrices to comprehend. Most students attempt to comply by reading policy and procedures manuals until their eyes glaze over. To be sure, it is important to become familiar with the policies of your organization; not doing so could lead to disaster. Really learning what an organization is about, however, requires more than just reading manuals. Later in this book we will discuss finding your place in the organization in greater depth. For now, it is important to remember that you are a newcomer to an already existing culture, and as such you must attempt to learn your way around. Ask yourself the following: Who are the players? Where do I fit into the structure? How can I begin to get acquainted? Where am I being asked to fit in, socially, physically, and professionally? Am I comfortable here? Is there anything I can do to carve out some personal space (particularly in a large, impersonal setting)? Who can help me learn the ropes about what I can and cannot do? Try to weigh the importance of every issue. For example, there may be a long waiting list for office space. Attempting to circumvent the informal code about waiting your turn could end up costing you more than an office is worth! Think about your style of introducing yourself and consider making an early effort to get around and actively acquaint yourself with at least some of your co-workers. If you typically wait for others to approach you, consider modifying your style. The result may surprise you and may influence how you approach similar situations in the future.

Meeting Your Supervisor

For students who have recently begun their fieldwork, or who are about to begin, one of the most critical events in getting started is the initial meeting with their supervisor. Most of you will approach this event with some trepidation. This person will be a central figure in your training, someone with whom you will work over the period of a semester or a year. He or she is charged with training you and evaluating your progress. It is understandable and reasonable for you to be somewhat nervous. Remember that although you cannot select the person who will be your supervisor, you can significantly influence the quality and content of the supervision you receive. Before meeting with your supervisor, or very early in the process, think about the specific goals that you have for the year. Prepare questions for your supervisor and don't be afraid to offer some ideas about experiences that you would like to have during your training. We suggest you construct a learning plan that outlines your goals and includes some activities designed to help you achieve those goals. Ask your supervisor about the range of activities available for you to explore, and don't be afraid to use some creativity in constructing the learning plan. If possible, talk to other students and staff to learn about ongoing activities in your setting that interest you. If constructing this kind of plan is already a standard requirement of your academic program or internship facility, this will be a relatively simple process. If it isn't, we advise you to initiate this planning on your own behalf. Many supervisors are busy professionals who will welcome your input. Keep in mind that the entire internship experience is dependent on your taking an *active* rather than a passive stance as learner.

VIGNETTE ONE

John is a 30-year-old graduate student in social work with a concentration in administration. The fieldwork setting is a multiservice community agency located in a metropolitan area near the university. This is John's first internship experience since beginning the program, although he has several years of experience as a manager in a non–social service field. He is a part-time student who plans to continue in his present employment as he completes his fieldwork. John is looking forward to a career as a manager in a social services or health care setting after graduation. At this point in his career, he believes that he has considerable experience in management, and he expects to have minimal difficulty adapting to this setting and his duties.

On the first day, John is struck by the hectic pace of the staff. When he tries to locate his supervisor, he discovers that he is only one of several other students and staff who are waiting for the supervisor's eventual arrival from morning meetings. When he does arrive, the supervisor is friendly but rushed as he points to a cluttered room with piles of paper, desks, and chairs (many in disrepair) and urges the students to assemble materials and supplies that they will need to get organized. He apologizes for being late and offers to be available "sometime tomorrow" if he can get through an emergency budget meeting scheduled just moments ago by the administrator. Feeling slightly irritated, John begins to assemble some supplies and notices a few of the other interns chatting nearby. He considers approaching them, then hesitates as they wander away together. After spending a few minutes looking around the building, John decides to return to the university in order to make more productive use of his time.

QUESTIONS FOR DISCUSSION

1. What might be some advantages/disadvantages of John's previous work history in dealing with problems as an intern in this setting?
2. What preparatory steps could John have taken to increase the likelihood of a successful first day?
3. How aware is John of the sources of his irritation? What alternatives exist for him to manage his feelings in an appropriate manner?
4. How might John go about expressing his personal reactions to the first day to his supervisor?
5. Based on initial information, what predictions can be made about this setting that will be useful to John for future planning?

ANALYSIS

It is not difficult to imagine the variety of reactions John might have to the typical experience of a first day at a new fieldwork site. Most likely, John arrived with a host of expectations about the new setting and a mental picture of how his first day would proceed. Interns anticipate their beginning days using information they have gathered from various sources, which might include an agency interview, a written description of the agency in a resource directory, a recommendation from an instructor, and experiences shared by other students who interned at the same setting. In addition to these acquired data, students construct ideas regarding a certain setting or supervisor based on previous encounters, accurate or inaccurate assumptions, and uniquely individual patterns of thought.

John arrived for his first day with many expectations about the agency, staff, and supervisor. (Remember, the more aware one is of his own expectations, the more he will be able to apprehend his reactions to the events that occur.) Let's consider John's mental representation of the setting. Did he anticipate the hectic atmosphere of the site? Perhaps he expected to find a supervisor or other staff person ready to orient him to his new surroundings. He may not have considered the possibility that he would be sharing his supervisor with a group of other interns. He may have imagined an orderly introduction to early assignments or a physical space designated for his use. Because John is a seasoned manager himself, he may have anticipated that his supervisor would provide an experience similar to that which he provides for new employees. He may have been unprepared to encounter significant age differences in other interns, staff members, or his supervisor. Given his age and experience, he may have predicted an easy adjustment to his new surroundings, or perhaps he surmised that he would be more comfortable than those with a more limited work history.

Deliberate and focused attention to your own thoughts and feelings may well contribute to increased understanding and management of early reactions during your first days of fieldwork. One word of caution, though: Don't allow your first impressions to carry undue weight! Early reactions and evaluations are influenced by an array of factors that will likely reduce their accuracy. Viewing a particular person or aspect of the setting in a fixed way based on limited information or exposure makes it difficult to assimilate new data that would result in a more realistic perspective. After all, you probably are hoping that your supervisor and other staff will refrain from making a hasty judgment about you and your abilities based on your first few days!

It may prove helpful to make one observation about John's hesitation to approach the other interns (who may well be sharing his first-day confusion). We can interpret his reaction to the other students in several ways. For example, is John's realization that he is one of a group of interns a surprise to him? Is he relieved to discover that there are others with whom he can learn, problem solve, and commiserate? Perhaps John perceives himself to be in competition with the others; this could add to his state of anxiety and irritation. Whatever the reason, John elects to delay introducing himself and perhaps forfeits the opportunity to experience a few gratifying moments in what seems to be a disappointing day. His hasty departure also suggests that avoidance may be a familiar response to discomfort. Although we all can empathize with the wish to indulge our urge to flee when certain unexpected complications arise, it often is wise to review other, more constructive responses.

VIGNETTE TWO

You are a counseling undergraduate with a strong academic background but little direct experience working with people. As a new intern in a clinic that serves children and families, you are aware most of all of your overall anxiety about starting your internship. In your initial meetings with your supervisor, you are relieved to find that she is easygoing and relaxed in her approach. You discover, however, that she expects undergraduates to carry the same workload as graduate students. Your tension level mounts as you listen to descriptions of tasks that sound completely overwhelming. When you begin to express some initial concerns, your supervisor is interrupted by a phone call. Later you defer all comment on the subject, preferring not to appear overly insecure to your supervisor.

A little later in the training phase of your internship, you find yourself having lunch with a few of the staff in the staff lounge. When one of them asks how things are going, you use the opportunity to share your initial reaction to your supervisor and to her decision to assign heavy workloads to all students. To your surprise, one of the professional staff begins to share her negative view of the same supervisor. It's a relief to discover a staff member who seems to understand why you are upset, but it makes you wonder if things might get worse as the semester wears on. Later, as you drive home, you find yourself wondering how many other staff members share your impression of your supervisor.

QUESTIONS FOR DISCUSSION

1. In this situation, which is more likely: the student is overreacting to the supervisor, or the supervisor is being insensitive to the student?
2. Was the student correct in withholding comment during the initial meeting in order to diminish the risk of being perceived negatively? At what point should the student comment on an item like workload expectations?
3. Do you foresee any problems arising as a consequence of the discussion in the staff lounge?
4. What aspects of the supervisor-supervisee relationship are appropriate for a student to share with other staff or students?
5. Should the student consider approaching his academic advisor for consultation prior to negotiating expectations with the on-site supervisor?

ANALYSIS

The initial meeting between supervisor and student is a little like a blind date. Both participants have their own expectations, which may or may not be realistic and which must be subject to modification. It is not uncommon for a student to have a strong reaction to a supervisor, nor is it uncommon for a supervisor to react to a student's initial presentation. There is no prototypical supervisor any more than there is one accurate profile of a typical student. As a result, both parties need to have a clear notion of their goals and objectives, and a willingness to be flexible. As a student in the above scenario, you face a variety of issues, any of which may require you to problem solve over a period of several meetings. It is important to consider the particular issues involved, apart from the interactive (relationship) process. For example, what is the appropriate level or amount of work assigned to a graduate versus an undergraduate student? What are your own work capabilities at the outset of an internship? In some situations, supervisors may spend a substantial amount of time discussing a student's background and training prior to assigning specific responsibilities. A student in one setting may have a dramatically different initial experience than his peer at another facility.

Consider the experiences you have had with a number of different academic instructors. Clearly, they are not identical in approach or philosophy, even when teaching similar course content. Students generally have strong initial reactions to instructors. It sometimes is not easy to separate one's initial response to workload or expectations from one's feelings about the supervisor as a person. Supervisors are human, too, and may be tempted to form broad impressions about an intern based on these early meetings. Memories of previous significant relationships can lead to bias or distortion when first impressions are forming. It's important for both sides to resist forming intractable impressions based on too little information, especially in this emotionally charged climate.

This vignette described a typical student interaction with professional colleagues. It illustrated how student experiences often are colored by ambiguity. Students are neither paid professional staff (even if they do receive a stipend), nor are they fully outside the professional ranks. The student subculture in the worksite is one that is distinct from the paid staff; there may even be more than one student subculture, depending on the size of the program. Student interns need to be sensitive to professional boundaries and become sophisticated about how to handle personal reactions to and conflicts with co-workers, other students, and their supervisor.

You may or may not question the appropriateness or wisdom of the disclosures made by the student in this vignette to his co-workers. Regardless of your reaction, though, the scenario described is fairly typical. This student, like most students, is seeking affiliation, support, and ventilation among peers and colleagues. Later on, the student may be faced with the consequences of certain choices. Many times choices are not inherently good or bad; any particular choice may lead to a desirable outcome. It is helpful to consider the possible consequences had the student made different choices—without judging any of the courses of action as either good or bad.

VIGNETTE TWO (VARIATION)

After your initial meeting with your supervisor you feel encouraged. It appears that your supervisor appreciates your input, has a clear plan to involve you in a hands-on training program over the course of the year, and foresees great progress for you in skill development. You leave the initial meeting feeling hopeful about the progress that you will make. Like most students, you depart this first encounter with few specific tasks, but you look forward to gaining greater clarity in subsequent meetings.

The following week you and your supervisor continue the discussion you began in the first meeting. Your supervisor presents you with additional information about your setting, the larger administrative structure, funding resources, and the meeting draws to a close. When you ask what specific tasks you might undertake in the following week, the supervisor instructs you to continue observing agency activities, meeting other staff, and familiarizing yourself with the setting. By the end of week number 2, your patience is beginning to wear thin and you find yourself reading class material and perusing magazines, newspapers, and staff memorandums. You look forward to the next meeting as an opportunity to establish some direction and secure some specific case assignments. The meeting is cancelled due to your supervisor's illness. No specific message is left for you about how to proceed. This frustrates you, but you anticipate some clarification later in the week. Wanting to be helpful, you write your supervisor a note indicating your availability to meet later if she is free. You stuff the message into her overflowing mailbox.

The remainder of the week passes and your initial anxiety is turning into irritation. The next week arrives and you finally are able to meet with your supervisor. During the course of the meeting you decide to assert yourself and bring up the topic of clinical assignments, as well as your concern about meeting class expectations for direct experience. To your bewilderment, your supervisor tells you that it would be to your advantage as a beginner to spend some time training alongside the clerical staff. She explains that by doing so you will gain valuable information about the inner workings of the agency, become better acquainted with support staff, and begin to contribute to the agency operation. Glumly, you accept the assignment, not wanting to contradict your supervisor. Before leaving the office you again express your hope to have direct practice assignments, and your supervisor assures you that they will be made available in the near future. Several weeks pass and you find yourself filling in for support staff, typing, answering phones, and filing medical records. You've heard nothing about any possible clinical assignments.

QUESTIONS FOR DISCUSSION

1. What are some appropriate activities for the beginning fieldwork student?
2. Should interns be able to decline assignments from supervisors? If so, under what circumstances?
3. Should a student ever be asked to perform support-staff activities (secretarial, clerical, and so on) as a percentage of his or her fieldwork hours? Why or why not?
4. What expectations for direct practice are appropriate for inexperienced students?
5. To what extent can a student expect to negotiate the terms of fieldwork duties?

ANALYSIS

This vignette presents a student who is attempting to cooperate with the supervisor and the agency while looking forward to a time when his own goals can begin to be met. A student in this situation may be tempted to act just as the example indicates, or he or she might pursue a more assertive course, challenging the supervisor about the duties assigned. (A new employee may well face similar problems. However, as a paid staff member, he or she may feel less uneasy about challenging or questioning assigned duties.)

There are several points to consider. The start-up period is the time when expectations are expressed, guidelines are established, and a course of action is defined. The student in this vignette may be having difficulty distinguishing between clerical duties that can legitimately be construed as "training" and those that basically constitute free secretarial service for the agency. An intern in this predicament may wonder whether the tasks are only temporary and therefore not worth complaining about. The same student might complain to the department or agency administrator or ask the university fieldwork advisor to intervene. Each course of action implies a different rationale and is likely to lead to a unique outcome.

Worth noting in this vignette are the student's strong reactions to ongoing events. Consider how you might react in a similar predicament. While you are contemplating solutions, you still must function in day-to-day activities involving school, family life, and so on, and you must continue to interact with staff and clients despite the obvious strain. The implication may be to pursue a solution as early as possible. The student in this vignette had reasons to anticipate a positive experience based upon initial meetings with his supervisor. Later, it became progressively more difficult for him to influence the situation.

Sometimes, the timing of interventions is as critical as the actions themselves. Students often are confronted with the need to take some action without knowing which action is the correct one. This is especially true when students are new to a setting and are feeling insecure. In such a situation, consulting with others may be the best way to explore and consider all the different options. In addition, the notion of establishing career goals and being able and willing to express them has application here. Candidates for internship settings who have an opportunity to conduct preplacement interviews may head off potential problems and/or identify sites that do not meet their requirements. Interviews also provide an opportunity for candidates to share their needs and concerns and to convey information about any prior experiences that may have been unsatisfactory. A supervisor may not realize a student's needs unless and until they are clearly conveyed. It is the joint responsibility of both parties to communicate.

SUGGESTED READINGS

Blake, R., & Peterman, P. (1985). *Social work field instruction: The undergraduate experience.* New York: University Press of America.

Bogo, M., & Globerman, J. (1995). Creating effective university-field partnerships: An analysis of two inter-organization models for field education. *Journal of Teaching in Social Work, 11,* 177–192.

Holtzman, R. F., & Raskin, M. S. (1989). Why placements fail: Study results. *The Clinical Supervisor, 6,* 123–136.

May, L. I., & Kilpatrick, A. C. (1989). Stress of self-awareness in clinical practice: Are students prepared? In M. Raskin (Ed.), *Empirical studies in field instruction.* New York: Haworth.

Nelson, D. L. (1990). Adjusting to a new organization: Easing the transition from outsider to insider. *Prevention in Human Services, 8*(1), 61–86.

Nelson-Jones, R. (1990). *Human relationships: A skills approach.* Pacific Grove, CA: Brooks/Cole.

Urbanowski, M. L., & Dwyer, M. M. (1988). *Learning through field instruction.* Milwaukee, WI: Family Service of America.

Chapter Two

Finding Your Place in the Organization

- Securing Your Personal Space
- Learning about the System
- Stretching Your Confidence and Expanding Your Capabilities
- Perceiving Yourself as a Member of a Network
- Vignettes
- Suggested Readings

SECURING YOUR PERSONAL SPACE

All new interns ask themselves the question, "Where do I fit in?" As your fieldwork experience progresses, it will be important for you to establish your own niche in the environment. This will be an ongoing process, not an isolated event. The degree of support that you'll receive will depend on the nature of your particular placement. Larger, more complex settings require a longer period of adjustment. However, such settings often have more structured training programs, which help define each student's role.

The first thing you should focus on is your physical workspace. Many students learn to cope with a work setting that offers little beyond the basic requirements. Still, it is important to be able to walk through the door in the morning and travel down the hallway to your own "office," even if that office consists of a small table, lamp, and a shared phone. A private office space, as desirable as it may be, is not essential to your survival. You'll be able to function adequately if you can secure a place that allows you some sense of permanence. (Being asked to constantly relocate will make you feel like a visitor who doesn't belong.) A locked drawer that houses supplies and belongings, and a desk, table, or shared workspace that can be scheduled for your personal use on specific days or times is a start. You can brighten a spartan or dreary environment by bringing in a few plants, photos, or personal items. When budgets permit, a door or desk sign with your name and title helps to create a professional designation for you as an identified member of the organization. (Business cards also serve this purpose.) Explore your options, but recognize that budgetary restrictions may limit your choices. It is advisable though to consult with your supervisor about intended plans for changes, moves, or additions in order to head off unintended conflicts with agency policies or co-workers.

Sometimes, staff or supervisors simply haven't considered alternatives; the creative ingenuity and thoughtfulness of students who redesign their work environment may well be appreciated.

Learning about the System

Professionals can often be heard describing themselves as being at the mercy of administrators, policies, and structures that limit their ability to act and perform duties in the most effective manner. Staff members may feel that their actions and opinions have no significant impact on the day-to-day procedures, policies, and activities that affect them. The reasons for this are numerous and are the subject of considerable study; however, the analysis of organizational behavior is a complex topic, one we cannot fully address in this text. As a new member of your fieldwork setting, though, you should try to develop an informed view of the systems that surround you and that you in turn influence.

One way to learn about the internship setting is, of course, to speak to other staff and students. Professionals who have been a part of the setting for a number of months or years have a wealth of information about everything from agency policy and work strategy to where to go for lunch. Of course, the information shared does not necessarily fit everyone or every situation, but knowing what other experienced staff know is always advantageous. You'll also want to find an opportunity to observe how other members of your profession are regarded in this setting. Is there a feeling of equality among those in similar disciplines? Does one group seem to dominate, while members of other groups are reluctant to participate and contribute? Sometimes close observation at a few staff meetings will provide clues about the relationships between different professional groups. It's a good idea to ask other students whether their observations are similar to your own.

Approaching a senior staff member for help or input demonstrates your openness. Most professionals enjoy the feeling of respect they gain from having their knowledge and training recognized. And, establishing working relationships with other members of your setting is positive for a variety of reasons. It certainly is more pleasant to feel you are an involved member of your professional setting than it is to feel disengaged and isolated. Also, accomplishing assignments over the course of the year will be much easier if you have access to help, support, information, and resources. Often it helps to identify individuals who can serve as models of competency, expertise, or achievement. Although no one person is likely to exhibit all the traits that a student admires, there is a place for the mentor or model in the development of each student who undertakes training in a complex discipline. A professional who demonstrates some of the qualities, skills, values, ethics, and level of expertise that you value can serve as a sounding board, advisor, or continuing reminder and example of what you hope to achieve.

It is important to remember that your own setting may not be similar to other settings in critical ways. This is why classroom discussion with other students can be extremely valuable. Comparisons and contrasts with other students can help to expand and define your impressions of the profession. For example, does the culture allow or encourage employee and student input? You'll also want to examine the role of supervisors in the larger structure. Are they allowed access to decision-making forums, or are they isolated? Does the staff have the opportunity to solve problems and implement changes, or does the organization discourage independent actions without approval from above?

An unavoidable question that all students have as they become familiar with working members of their profession is, Do they enjoy their work? If not, why not? Is the problem related to the work site or the person? Is the individual miscast in his or her position? Have I harbored any misconceptions about what members of the profession actually do in day-to-day activity? Although there is no simple method for answering these questions, we encourage you to gather information from a variety of sources, without hurrying toward a premature conclusion.

The examination of multiple sources of information about these issues allows for a fair and objective appraisal. Sometimes the opinions of employees reflect their individual bias or a viewpoint that is more emotional than scientific. Similarly, the examination of too few events (for example, judging the social climate on the basis of one or two staff meetings) can lead to an overgeneralization.

Develop an Active Stance

Once you have begun to develop a working knowledge of the basics, you can consider how to begin and maintain an active input to the organization. As a new student, you have the opportunity to be active in your setting rather than passive, inquisitive rather than accepting, creative rather than resigned. This does not imply that you'll be able to implement changes simply or without strategy or effort. It does imply that in order to maintain some control over your fieldwork experience, you'll need to invest some energy in understanding the organization as a means of determining your own course of action. This may be easy or difficult depending on the setting. Some settings are more friendly as a culture than others and encourage students to take an active role. Certain departments and department administrators are more flexible than others. Some administrators or supervisors may prefer students to remain silent and accept instruction gratefully; such a style may suffice when instruction is in keeping with student interests. Usually, however, such an approach deprives students of the opportunity to learn to be assertive and to grow in confidence and competence. It is always more desirable that a student learn to assume an appropriately active and contributory role, rather than a passive one.

Recognize Your Own Value

It is easy to overlook your contributions to the setting when you are a beginner, especially when you are uncertain of your own specific value. It may be helpful at this point to recall that even as a novice in your chosen profession, you have much to offer. Every student brings valuable life experience, related expertise, training, and some degree of exposure to problem-solving strategies. Early in fieldwork, prior experience may seem to be of little help or relevance. As time passes, it will be easier to integrate prior knowledge with current internship demands.

Remember, students have a reciprocal relationship with their training facility. In some instances, the agency or program simply could not exist without the students. More commonly, the student population contributes in a variety of ways to workload production, providing direct services to clients, completing intakes, and generating increased revenue. Other intern activities may include producing reports, analyzing data, assisting in fundraising, planning community events, organizing committees, participating in marketing, and in multiple other ways supporting the daily functioning of the setting.

Analyze and Influence the Quality of Your Training

How your organization views student interns is a critical factor in determining the degree of fit and comfort you can achieve in your setting. As you begin your training, you'll want to take note of any programs or services that are designed to help teach and prepare inexperienced interns and expose them to senior staff. For example, is there a weekly or monthly forum for discussion of difficult cases, agency projects, or research in progress? Are students expected (or allowed) to participate? Are students active in staff meetings or committees? Or is the administration's attitude best described as laissez-faire?

Some facilities offer training that is varied and multifaceted. For example, large training hospitals often have specific programs in place, coordinated by an individual who is responsible for overall quality. Lectures, films, discussions, videotape presentations, case presentations, and the like all have a place in an effective training program. For those students who are fortunate, many or all of these features may already exist at the time that they begin their training. Newer settings that are unfamiliar with having students on board may be less adept at integrating students into ongoing activities and may place less priority on training. In every instance, student input can help identify the interest and need for training and may provide the impetus to begin a program or to enhance an existing model. (Sometimes students might discover that the training is actually targeted to an allied discipline to the exclusion of their own group. Here the assistance of the fieldwork supervisor can help.) Other settings will fall short of elaborate or even adequate training. When several students or students and staff collaborate in identifying the need for training, the likelihood of success grows. Groups who work in collaboration often exert greater influence than a single group, particularly one with the limited power and status of the student population.

Collaborate with Other Students

We've stressed the need for students to assume an active and creative role in their professional training. Some of the most productive and enjoyable collaborations occur among students. Many times, students are best able to institute creative and enjoyable activities in their fieldwork site after talking with one another about areas of mutual interest. It is not uncommon for students to plan specialized therapy groups, research projects, agency redevelopment plans, and so on, as an outgrowth of conversations following a long day of internship. Networking with others allows students to vent about a day's (or week's) worth of frustration or accumulated stress. Even the best fieldwork settings aren't perfect, and even the best students make mistakes. Sharing their experiences with others enhances students' self-esteem and sense of belonging. Taking the time to talk with other interns about your experiences and frustrations will allow you to discover that you are not alone. The stresses of juggling work, classroom, and assignments often become overwhelming for even the most talented and capable student. Establishing a network of friendships among student colleagues may be one of the most significant factors in making your fieldwork experience a successful one. Remember, the burden falls on you to initiate and foster ongoing connections with other students who share your setting. Make sure you find opportunities to network and participate with students of allied or related disciplines; your training and responsibilities probably overlap. Assuming a too rigid boundary between yourself and the students from other schools or programs is likely to close avenues for creativity, learning, and friendship.

Stretching Your Confidence and Expanding Your Capabilities

Over time, every student and professional develops a reputation in the agency or institution for their ability or failure to contribute. How these perceptions are formed is interesting to consider. Students or other professionals who are perceived as competent and confident often convey a willingness to take on difficult or complex tasks. They consistently demonstrate their ability to problem solve, manage stress, and relate well to others. These individuals are seen as growing, developing, learning, and interested in their activities. Remember, professional communities are small, and students who are active, valued, and confident may receive employment opportunities later as one consequence of a strong performance.

Students (or employees) who are consistently reticent or take on only those duties that fall within a narrow, predetermined range are eventually labeled as limited. Unfortunately, professional reputations, once established, are enduring, and follow individuals from setting to setting. The reputation you establish as a student may become synonymous with your professional identity.

Develop a Professional Identity

Your professional identity is a reflection of your overall competence, level of technical skill, and value to your organization and its clients. It likely will change several times over the course of your career as you continue to learn and improve, enhancing your fund of knowledge, your sophistication, and your skill level. Typically, professionals take on increased responsibilities over time; a progression of duties reflect their growth as employees. Senior staff members often enjoy an enhanced status within the organization; they may even be assigned formal titles such as supervisor, trainer, or senior consultant to junior staff or students. All such designations contribute to their sense of value, status, and esteem among colleagues. All help to define their professional identities.

Students, like staff, begin to establish their professional identities from the very beginning of their fieldwork. Students who have had considerable work experience may find it easier to negotiate the internship setting and may enjoy a feeling of competence and a sense of contribution early on. Students in administrative disciplines who previously have held positions of authority may find it natural to view themselves as managers or administrators, even at the outset of their internship. In contrast, inexperienced counseling interns facing their first interview with a family may worry that their inexperience or anxiety will betray the uncertainty they have about assuming an unfamiliar role. Classroom training is not the same as actual experience; time and supervised practice are needed for the intern to integrate academic training with applications in the field. Where there is a considerable difference between one's self-perception and the role assigned, difficulty arises and there is a need for support. Once again, students' need for involvement with supervisors, instructors, and experienced staff in order to negotiate this challenging territory is illustrated. The student may discover that senior staff have valuable perspectives based upon their own familiarity with the very same struggles. The mere discovery of not being alone in this area is typically a source of considerable relief.

The mere factor of experience, however, does not ensure greater proficiency or easier adjustment. Students with considerable previous practice may actually experience a greater degree of stress than their more unsophisticated counterparts. Imagine a situation where a younger supervisor or manager oversees the performance of a student who has held several positions of considerable authority. In this setting and despite previous successes, the student probably will not enjoy the status, power, authority, or respect that he had in his former work settings. Perhaps the student will bring a remarkable degree of maturity and self-assuredness to the experience and will undertake the challenges with considerable grace. More likely, he will have to struggle to adjust to the present arrangement. Of course, every student is different in personality, skill, and level of confidence. All are human beings subject to some expectable trials and tribulations. Students may be tempted to cope by elevating themselves to a position similar to "honorary staff member." Generally, other staff and students are aware of a student who is attempting to distance himself from the student population and align himself with the staff. This strategy may result in a temporary "elevation" in status, but the cost may be a loss of connections with student peers. The student may discover that he is neither a member of the staff nor of the student subgroup.

Difficulty in establishing a professional identity also can occur when a student is unable or unwilling to assume the responsibility of his or her position. The intern may be overly anxious, constantly asking for clarification or assistance with even the most minute or elementary task. Over time, it becomes apparent that the individual is not ready to take on the usual tasks, such as organizing a meeting, networking with colleagues, speaking to a community group, presenting a report, meeting with clients, and so on. Initial reluctance is not a sign of poor fit or lack of ability. Continuous or excessive withdrawal from activity and an overreliance on others is a sign of difficulty that suggests the need for consult with the supervisor or academic advisor before problems become too entrenched.

Accept the Challenge and Struggle of Learning

Any serious pursuit of learning involves struggle and difficulty. Inevitably, some errors will be made along the way. Supervisors in the fieldwork setting expect this and anticipate that students will encounter some situations with ease and others with a feeling of confusion and uncertainty.

One factor that often contributes to students' anxiety is the pressure they place on themselves to perform at a high level. Many students with a history of significant accomplishments are so accustomed to success that the idea of performing less than perfectly is intolerable. Undergraduate and graduate students alike can be fiercely competitive with one another and have little tolerance for below-average performance. Interns must remember that they are there to learn; they must allow themselves to act without excessive fear of failure.

Supervisors are accustomed to the idea that students may downplay their mistakes or attempt to shield supervisors from their problems. This is understandable; students worry about how mistakes will affect their performance evaluation or access to future assignments. Without the opportunity to work on problems, though, both the supervisor and the student are deprived of opportunities to learn. Students must be willing to describe situations that they need assistance with at any and all times in their training. If they fail to do so, further damage may result. We encourage students to initiate a general discussion about this important topic early in their fieldwork training.

Perceiving Yourself as a Member of a Network

Supervisors are not the only source of instruction and support in the fieldwork setting. Professionals, generalist human services workers, support staff, administrators, student colleagues, and others are all potential sources of learning. As a student, you will do yourself a disservice by taking a rigid view of the workplace that limits your opportunities in any manner. For example, a student colleague with whom you have had little chance to become acquainted may turn out to be both informed and well connected, and may welcome an opportunity to assist a fellow student. You may be a valuable source of help to a colleague or staff member. Staff and students can and do help one another on a regular basis. Students, clients, and all receivers of care or service engage in give-and-take with the organization and its members.

One of the dilemmas of internship is that most interns (with some exceptions) provide services without financial compensation, creating an extra stress that most students could live without. If you are not one of the lucky few to receive a paycheck along with your experience, you'll have to take the view that the intangible rewards count even more than the monetary ones! Without an appreciation of the future rewards of your training, intern life can be wearing and difficult. Remember, many students often look back on the relationships formed in fieldwork as the most valued and enjoyable aspect of their experience.

Vignette One

A social work student interning in a large, urban teaching hospital settles into the routine of seeing clients, running groups, and attending meetings. In this particular setting, social workers commonly are utilized for family and group therapy. This student hopes to develop skills that cover the entire spectrum of treatment, including diagnosis and individual therapy. During the course of the semester, the supervisor informs the student that a series of weekly training sessions is offered for psychiatric residents at the main hospital building. The training sessions address a variety of mental health topics, most of which are in the student's area of interest. The supervisor advises the student to attend an upcoming meeting on group treatment (an area of special interest to the supervisor). Future training topics include clinical diagnosis, uses of psychiatric medications, and individual therapy—all regular topics of training for physicians in psychiatric residency.

At the group treatment training session, the student notices a few other social workers, along with nursing and clinical psychology students. When the student expresses interest in attending future training sessions, the supervisor appears less than enthusiastic. Citing the fact that the training is funded by the medical department for psychiatry residents, he discourages the student from weekly attendance, suggesting that only a few additional trainings may be advisable in order to avoid "irritating anyone." The student considers attending the trainings secretly.

QUESTIONS FOR DISCUSSION

1. Do students have a right to expect access to interdisciplinary training? What are the advantages of training in an interdisciplinary setting? What are the disadvantages?
2. Should this student have considered approaching the training coordinator directly?
3. What should the student do if he or she is not allowed to participate in other training sessions?
4. What factors may be contributing to the supervisor's position?
5. What actions by the student may have the effect of increasing or decreasing tensions between the respective departments?

ANALYSIS

The student in this vignette is interning at a facility that houses multiple disciplines, many of whom share and compete for resources and space. (The disciplines may overlap in areas of interest and responsibility as well.) The vignette suggests many issues that may arise in such situations. The student may view himself as clearly belonging in an area where the supervisor feels less comfortable. However, areas that may be of professional interest to the student may seem remote or out of bounds to the supervisor. In that instance, the student and the supervisor have some work to do defining scope of practice.

There may be reasons that explain why the training opportunities in the medical department should be cautiously utilized. For example, there may be a history of excluding groups other than medical residents from the training. Similarly, there may not be avenues for the reciprocal training of physicians via departments of psychology or social services.

Any organization will have its own methods for utilizing staff members and encouraging or limiting flexibility in job tasks. For example, a hospital may draw stricter boundaries between physicians and social workers than a community clinic. Of course, all hospitals and all clinics are not identical, and roles, organizational hierarchy, and overall structure will differ depending on the site. Learning about the specific structure in your fieldwork placement is a must; trainings or staff meetings will provide opportunities for you to discover the interpersonal and organizational dynamics that are present in each new setting you encounter.

Students often become interested in activities that occur among other staff and students both within and outside of their own departments. This is natural; professionals often are called on to perform tasks that involve collaboration with others and require a process of ongoing learning. In order to converse with members of allied professions, health professionals may need to learn the terms or jargon typically associated with that group. Medical records, for example, may reflect ideas from multiple disciplines, depending on who is involved in the case. Professionals also must be able to collaborate effectively with one another when working with a patient or client. Sometimes, areas of responsibility may become blurred or confused. Eventually, a professional may realize his or her skills extend far beyond the limits that were defined early in his or her training. The very nature of one's own discipline may even change, along with technological growth and the gradual expansion of the knowledge base of the profession. Assessment and adjustment are the rule rather than the exception.

VIGNETTE TWO

As a student in child development at a local elementary school, you are assigned a mixture of responsibilities that include classroom assistance, observation, activity planning, and interaction with school counseling staff. Over time, your usual activities become routine, and you begin to consider alternatives that may allow for a greater degree of creative input and provide the opportunity to play a lead role in a project. A classroom assignment requiring the construction of an original exercise leads you to design a workshop for parents. For several weeks you plan and develop ideas that can be used to formulate a program that involves faculty and parents, educating them about student needs and improving parent-teacher collaboration. Eventually your supervisor suggests that you consider discussing the ideas with counseling staff in order to gain their input. Together with the staff school counselor and a school psychology intern, you plan a parents' night that includes group discussion and presentations on topics related to classroom learning, child development, and management of problem behavior.

A date is selected for the event and information is sent to the families' homes to promote attendance. Unfortunately, the date selected conflicts with your own schedule at a very busy time of year. Indications are that the date is convenient for the school psychologist, who had voiced a strong preference. Flyers announcing the event were printed at school expense prior to any agreement on a date by the planning committee, making the possibility of changes unlikely due to cost restrictions in an already tight school budget. Feeling resigned to the decision, you alter your own plans and move forward with final preparations. On the evening of the program, you are pleased with the excellent turnout but are soon dismayed to find that the activity you had planned to lead has been cut from the agenda at the last moment by the school psychologist. In addition, the program format has been altered, eliminating your opportunity to participate in a second presentation that you helped develop. You feel disappointed and unappreciated and wish you had not volunteered so much of your valuable time.

QUESTIONS FOR DISCUSSION

1. How could the student have invited participation differently?
2. Could the student have done more to confirm the acceptance and implementation of her ideas?
3. What should the student have done when the original schedule became a problem?
4. Should the student have approached the psychologist when the original agenda was altered on the evening of the event?
5. What is the appropriate follow-up by the student subsequent to this experience?

ANALYSIS

The student in this vignette encounters a situation involving a number of players who occupy differing roles in the work setting. As is typical, these players do not necessarily have equivalent power or influence. As a result, the outcome of the situation is influenced by forces not fully anticipated by the student. A retracing of the events may suggest alternative strategies that the student might have found to be useful. Although the student may have the satisfaction of having significant creative input to the structure and development of the parents' night, it is apparent that she suffered a major disappointment. In even the most straightforward and simple organizations, unanticipated events will occur that must be accommodated. In this scenario, the significant players may not have been involved at the outset in a manner conducive to maximum cooperation. Also, there may have been a conflict of aims or intent among the participants that may have influenced how the plan unfolded. As is often the case, the relative influence of the student may have been diminished by more senior staff members. In retrospect, the planning, collaboration, and problem solving of the student can be reconsidered from a new perspective. What might have allowed for a better result?

VIGNETTE THREE

A nursing student in fieldwork at a university medical center is assigned to colead a support group for adult cancer patients. The group is a mixture of men and women patients, all currently receiving treatment. The student is joining an ongoing group that already includes two other leaders, an R.N. staff member and a medical resident. Although the experience is initially somewhat intimidating, the student gradually assumes a degree of comfort in leading the group, despite her lack of experience. After a few weeks the staff R.N. leaves the group with little warning in order to assume a new position.

As time passes, the intern becomes aware of some tension developing between herself and the medical resident over the direction the group is taking. She brings the matter to her supervisor for help. The supervisor advises her to speak to the resident directly to express her concerns. When the student does so, the resident dismisses her concerns as unimportant, suggesting that she lacks the experience to adequately judge the situation. He reminds the student that he has been a member of the group for a longer period of time than she has been. At the next staff meeting, the disagreement between student and resident becomes obvious. As a result, the student's supervisor is mildly criticized by the unit medical director for failing to adequately prepare the student prior to her entry into the group. The medical director then suggests that the resident spend some time with the student to help familiarize her with treatment protocol.

QUESTIONS FOR DISCUSSION

1. At what point should management intervene in staff-student disagreements?
2. Should staff and students be expected to collaborate in providing services?
3. What steps can be initiated by the supervisors and/or by the student to help facilitate the transition of the student into an existing worksite or project?
4. What might be the experience of the patients in the scenario described?

ANALYSIS

The student in this scenario experiences some of the events that may happen to any newcomer to an already established workgroup. The ease with which the various leaders interact with one another and carry out their tasks is influenced by a number of factors, such as the leaders' prior experience, job status, affiliation with management, and existing relationships. In analyzing this situation, you may want to consider the relative strength of each of these factors from the perspective of the student, resident, R.N., or medical director.

In real-world situations, planning and accurate foresight are sometimes missing. Staff and students may be assigned to tasks as various situations arise, without a full and complete consideration of the likely outcome. Alternately, mistakes may occur in spite of thoughtful planning. Identify whether adding or subtracting certain steps from this process might have a positive influence for the student, the resident, or the group participants.

SUGGESTED READINGS

Gelman, S. R. (1990). The crafting of fieldwork training agreements. *Journal of Social Work Education, 26*, 65–75.

Grossman, B., Levine-Jordano, N., & Shearer, P. (1991). Working with students' emotional reactions in the field: An educational framework. In D. Schneck, B. Grossman, & U. Glassman (Eds.), *Field education in social work: Contemporary issues and trends*. Dubuque, IA: Kendall-Hunt.

Handler, J. (1992). Dependency and discretion. In Y. Hasenfeld (Ed.), *Human services as complex organizations*. Newbury Park, CA: Sage.

Hasenfeld, Y. (1992). The nature of human service organizations. In Y. Hasenfeld (Ed.), *Human services as complex organizations*. Newbury Park, CA: Sage.

Neukrug, E. (1994). *Theory, practice, and trends in human services: An overview of an emerging profession*. Pacific Grove, CA: Brooks/Cole.

Schmid, H. (1992). Strategic and structural change in human service organizations: The role of the environment. *Administration in Social Work, 16*(3/4), 167–186.

Thibadeau, S. F., & Christian, W. P. (1985). Developing an effective practicum program at the human service agency. *The Behavior Therapist, 8*, 31–34.

Chapter Three

Making Use of Supervision

- What Is Supervision?
- The Developmental Stages of the Intern-Supervisor Relationship
- Developing the Supervisor-Intern Relationship
- Vignettes
- Suggested Readings

Through the ages and across cultures, novices in many crafts, arts, and sciences have sought to expand and cultivate their knowledge by serving as apprentices to seasoned practitioners. Such arrangements allow teaching to be tailored to the needs of the individual, so learning is maximized. The relationships are mutually beneficial and represent a unique method of instruction that contributes to the continuity of entire fields of work. In human services, the notion of fieldwork is valued as a means of providing students a laboratory in which to practice skills and integrate theoretical concepts. Students function alongside staff members, accessing practical information as they begin to develop their own identity as professionals. The supervisor-intern relationship is at the center of the fieldwork experience.

What Is Supervision?

As students undertake the process of locating possible fieldwork sites and arranging initial interviews to secure a placement, a primary consideration is the opportunity to obtain high-quality supervision during the internship experience. Many internships require students to be available for several hours each week to perform their various duties. Typically, these hours are given on a volunteer basis. What students obtain in exchange for hours (and in lieu of financial gain) is supervision.

Supervision is a method of training and teaching in which experienced professionals interact with students and interns to provide guidance, on-site education, skill development, and general support. Aspects of supervision often include direct observation of the student, training meetings arranged for the specific purpose of addressing the needs of staff members and interns, weekly individual or group sessions where interns discuss their work with their supervisors, and, periodically, goal-setting and evaluation sessions.

The form(s) of supervision that occurs in a given setting will be a function of many factors. Some fieldwork sites have well-established training programs that have been developed over a period of several years; these programs often are finely attuned to the needs of interns at various levels and represent a commitment to the development of emerging professionals. Some portion of their funding often is dependent on the continuing existence of the training program; substantial revenue may be generated by the interns. In other agencies, although interns frequently may be used to expand the range of available services, training efforts may occupy a less prominent role in the facility. Supervision experiences may take place informally, with only a moderate level of organization and consistency. In addition, some agencies may be in the beginning stages of constructing a training program. Although such sites generally are receptive to interns and attempt to provide an adequate experience, interns in these situations may become frustrated by predictable start-up problems, particularly when they are inexperienced themselves and in need of direction and guidance. Finally, some settings sponsor intern involvement but are lacking in knowledge or awareness of the training responsibility that is implicit in the intern-agency agreement. Students may find little support for their requests for supervision and may have difficulty scheduling consistent meetings. Often, the intern and supervisor do not share a common definition of their respective roles and the opportunity for a mutually beneficial relationship is lost.

Universities that sponsor fieldwork programs should make every effort to periodically review agencies approved as training sites in order to judge their overall caliber and their ability to provide an adequate training environment. Students also must take some responsibility in ensuring the quality of their training experiences. Knowing how to measure the suitability of an agency is fundamental to achieving success in a training program. It is unwise to assume that a given site will be satisfactory based solely on a listing in a directory of approved or suggested agencies. Program and personnel changes are common in the field of human services and contribute to substantial variations in the quality of training and supervision. On-site interviews or visits to various facilities prior to committing your time will be beneficial. It also may be helpful to contact students who were previously in placement at a site you are considering, and to focus on questions regarding the quality and availability of supervision.

Additional suggestions concerning the selection of fieldwork sites will be the subject of a later chapter.

The Developmental Stages of the Intern-Supervisor Relationship

Like many endeavors that are grounded in interaction with others, the supervisor-intern relationship is an evolving process. It involves a unique combination of professional and life experiences, personal qualities, similarities and differences. It requires an interweaving of teaching and learning styles and active communication on a regular basis. Each of the individuals involved brings strengths and weaknesses to the project of constructing a working relationship that will be of mutual benefit. Let's consider the developmental processes inherent in a typical supervisor-student relationship.

The Beginning Stages

In the beginning stages of the relationship, the intern may have a strong need for specific direction, for focused instruction regarding procedures and policies, and for assistance in developing suitable objectives. In the first few weeks, especially, interns depend on the reassurance and support of others, although the needs of particular interns may vary depending on such factors as age, maturity, prior experience, interpersonal skills, and so on. It is during this time that first impressions are formed, initial goals are set and the tone of the relationship begins to emerge. Clear and direct communication will facilitate a smooth start and promote an atmosphere of cooperation as the intern and supervisor discover how to work well together.

Understanding how you feel about playing the role of novice is very important. One student may be quite comfortable functioning under the direction and guidance of a more experienced professional, asking questions as the need arises and assertively engaging in discussions. Another individual may have a difficult time in the subordinate role due to personality factors, uneasiness about being the new intern, a reluctance to ask for help, or negative reactions to dealing with authority figures. Self-awareness in this area is critical. You need to be able to recognize familiar patterns of behavior as you begin to establish a relationship with your supervisor. All too often, students focus on what they deem to be the shortcomings or personality defects of their supervisors while failing to recognize their own contribution to difficult interactions. Consider the following questions:

1. How might age influence your experience of a supervisory relationship (your age, your supervisor's age, age differences)?
2. How might you react to a supervisor of the same gender? different gender?
3. What personality factors, if any, are important to you in a supervisor?
4. What effects might cultural differences have on the intern-supervisor relationship?
5. What might influence your ability to ask questions of your supervisor?
6. Under what circumstances might you hesitate to reveal your need for help or additional instruction?
7. In what ways do you feel dependent as a novice?
8. In what ways do you function independently as a novice?
9. How might you describe your previous experiences with authority figures?

The Middle Stage

The next stage of development in an intern-supervisor relationship typically is characterized by an increasing level of trust in the intern's ability to function with a moderate degree of autonomy. The intern is familiar with the ground rules of the setting and the expectations of the supervisor (if communication has been clear), and a routine of activity is to some degree in place. Each individual is becoming acquainted with the other's personality, work habits, communication style, and areas of interest. An active supervisor will constantly assess the skills of the intern, providing opportunities for more complex tasks as indicated. Interns should play an active role in this process as well, lobbying on their own behalf as they feel ready to assume new projects. Contracted goals may be negotiated many times during the course of an internship. Sometimes the supervisor and intern will agree on appropriate objectives or progress toward a specific goal; under these circumstances both parties will be satisfied with the outcome of the negotiation. At other times, interns will be expected to defer to the opinion of the supervisor or the policy of the agency as to what direction their training will take. (If a serious dispute arises and cannot be resolved, requesting the assistance of university-based fieldwork coordinators is a responsible step to take.)

No two interns are likely to progress in the same manner or pace, even in identical practicum settings. When you are in the position of intern, you must assume primary responsibility for monitoring your level of satisfaction. Track your progress toward goals and press for exposure to areas of special interest to you. Present your concerns in university fieldwork seminars, to your fieldwork instructor, to other interns, and to classmates. Ask for assistance in evaluating your placement and compare your experience with those of others in similar settings. Often interns have legitimate issues to raise with their supervisors but hesitate to discuss these matters directly. Consulting with others in your seminar can bring clarity to a given situation and, perhaps, provide an opportunity to plan an effective supervisory conference.

The Final Stages

The final stages of the intern-supervisor relationship occur during the final phase of the fieldwork experience. It is a time for advanced skill development, an increasing ability to be frank about needs and expectations, and guidance through the process of termination with clients and colleagues. There will be a period of closure, during which evaluations are completed. On the part of the intern, the desire for autonomy increases and a deeper level of confidence is manifested. The supervisor responds by permitting a larger measure of independent functioning and encouraging the intern to exercise sound judgment skills in the performance of duties. Interns interpret the supervisor's stance as deserved acknowledgment of their growth and recognition of their diligent work. Frustration develops when a supervisor's style does not allow for the inevitable consequence of good training: an intern who is much less dependent on a supervisor for constant direction!

As the internship is drawing to a close, both supervisor and intern have many tasks to complete. Termination is an exceedingly important time and there are several issues to consider during the weeks preceding the final day of work. When fieldwork has been a successful experience, interns will become even more aware of the deep investment they have in their work and their clients as they are preparing to depart. They may need to rely once again on supervisory guidance through what can be a complex process; they will make new discoveries about personal patterns that arise when they are faced with separations and endings. This also is a time of evaluation, during which intern and supervisor reflect on performance and level of satisfaction. Most universities have formal paperwork that asks the supervisor to rate the achievements and progress of the intern across a spectrum of skills and interpersonal abilities. Interns also may be asked to complete an evaluation of the agency, and to note the quality of their supervision and overall experience. It is helpful for the supervisor to provide the results of his or her evaluation to the intern, allowing for sufficient time to review and discuss ratings and reactions in a concluding supervisory conference. Interns welcome the opportunity to be included in the evaluation process and can supply valuable feedback to their supervisors that may result in changes to benefit incoming students.

Developing the Supervisor-Intern Relationship

Ideally, the search for a fieldwork placement includes a meeting with the individual who will function as your supervisor. We advise you to request an opportunity to introduce yourself and briefly discuss initial supervision concerns, which include the schedule of meetings required by your academic program (frequency and length) and the format of supervision usually provided by the fieldwork site (individual or group meetings). Exchanging information about learning and teaching styles and individual expectations is also important. In preparation for such an interview, you must define your interest in the internship and develop some specific learning objectives. It will be difficult for you to judge the suitability of a setting if you have not clearly determined what you wish to learn. Focus on the skills you would like to develop or enhance and the activities (including the level of involvement) you would like to pursue; ask the supervisor if these experiences would be available to you as an intern. Engaging in this process will enable you to make an informed decision about a fieldwork placement and will increase the likelihood of satisfaction during your stay.

Contracting to Meet Your Needs

As you then begin your fieldwork hours, one of the first important tasks you will face is developing a contract that defines learning objectives, the manner in which you will achieve those objectives, and the various methods of evaluation to be used by the supervisor. This task provides an opportunity for you to strengthen your relationship with your supervisor; strive for reciprocal interaction and shared responsibility. Formulate *specific* ideas about goals and activities and be prepared to enter into a discussion with your supervisor. Take an active role in designing your fieldwork experience; it is a sign of your developing professionalism and will be appreciated by busy supervisors. Remain open to suggestions and ask clarifying questions. If the supervisor asks you to undertake tasks that do not seem immediately relevant to your conception of the internship, ask how these tasks relate to your learning objectives. Although you may not always understand or concur with a particular policy or agency requirement, a candid conversation with your supervisor may reveal the logic behind a given procedure. As a beginner in a new setting, you are not likely to be familiar with the historical perspective in which policies are considered; requesting more information permits a broader view that often leads to greater flexibility and tolerance.

Your completed contract will be comprised of several realistic goals that represent the combined needs, expectations, and areas of interest of your particular intern-supervisor dyad. The document may be renegotiated at various times during your internship by using the same format of interactive discussion. A working contract defines the expectations of both parties. Objectives are specific and arranged in a progressive manner, reflecting activities that are geared to an appropriate level of skill development and are designed to promote learning. The contract describes the frequency, length, and time of scheduled supervisory meetings. (Be sure to obtain a firm schedule of meetings; lack of a predictable and reliable framework for supervision is one of the leading causes of frustration for interns.) Because some supervisors and fieldwork sites approach the concept of supervision in a less formal manner, you may wish to present a description of your academic requirements to your supervisor.

Creating a Working Alliance

In the initial weeks of a practicum experience, interns and supervisors gradually become aware of similarities and differences in their approaches to issues of professional practice and training. Although students may hope for a supervisor who will be readily available and intuitively aware of their questions and needs, this probably is not a realistic expectation. Supervisors may wear a number of hats and function in many roles in the course of an average day. It is important to continually remind yourself that this partnership will function well only if both parties make an earnest effort.

Here is a partial list of familiar intern complaints:

"I can never find my supervisor when I need him."

"My supervisor seems to have so much to do that I feel I am imposing on her time."

"Although my supervisor is a likeable person, I find that I'm not learning very much."

"My supervisor has so many years of experience that he's forgotten what it's like to be an intern."

"My supervisor doesn't seem to appreciate how hard I work. She never thanks me or makes a positive comment."

"I thrive in situations that are well organized and structured. Although I admire my supervisor's flexibility and creativity, I often feel frustrated and lost."

"My supervisor and my son are nearly the same age, and I feel uncomfortable taking direction from her."

"I'm surprised that my supervisor didn't know I would need more guidance before trying this on my own."

"I think my supervisor expects me to interview clients exactly the way he does."

In fieldwork seminars, students commonly present and explore difficulties like those reflected in the above statements. Shortcomings in communication can give rise to a variety of misunderstandings and faulty assumptions that undermine the intern-supervisor relationship and impede its effectiveness. It can be very helpful to discuss your concerns with others to develop greater clarity about a given reaction or situation. Try to understand *your* part of the interaction before focusing too much attention on your supervisor. It is not productive to expend a great deal of energy guessing what might be motivating your supervisor to behave in a certain way; even if you could correctly guess the nature of your supervisor's behavior, you still would be faced with a set of circumstances that require an adaptive response on your part. You will have devoted valuable time to a pursuit that aids you very little in coping with the issue at hand.

Good communication requires the ability to express ideas clearly and to have an open mind while listening to others. Once you have gained sufficient understanding of an issue or reaction that is causing you discomfort, try to find a suitable time to discuss it with your supervisor. Initiating such an exchange may pose a challenge, depending on your mastery of certain interpersonal skills, including assertive communication, conflict resolution, and the capacity to negotiate. Role-playing a potential interaction in a fieldwork seminar allows you to rehearse alternative approaches and anticipate any problems. Students often feel anxious about participating in these exercises before a group of their peers, in part because of the improvisational nature of the activity. Concerns about being judged by others and reluctance to reveal areas of struggle are quite common and can be alleviated somewhat through discussion prior to the role-play. Remember that interpersonal skills improve dramatically with practice, making any temporary discomfort seem worth it. Identify those skills you may need assistance with and initiate a practice session in your fieldwork or practicum seminar.

Exposing Problem Areas

Judging by our dialogues with interns in academic and fieldwork settings, two particular issues seem to arise frequently in regard to creating a working alliance with supervisors. First, interns often are quite concerned about exposing problem areas to their supervisors. Knowing that the supervisor functions simultaneously as teacher and evaluator can result in a reluctance to disclose areas that may need attention. The desire to be seen as a competent human services worker competes with the need to recognize and reveal gaps in knowledge and experience. An intern's self-confidence may be fragile, and acknowledging mistakes or the need for help may seem threatening.

At the core of every good supervision experience is the capacity of intern and supervisor to develop an atmosphere that invites open discussion of both successes and failures, strengths and weaknesses. This is a collaborative process involving the supervisor's ability to be aware of and sensitive to the issues common to interns. Supervisors have a responsibility to explore any aspect of the intern's performance that affects overall learning. At times, this requires the supervisor to identify a problem area of which the intern may or may not be aware. Success will be determined largely by whether the supervisor takes a respectful and nonjudgmental approach. Supervisors will vary as to interpersonal manner when confronting or criticizing performance issues; therefore, a discussion of teaching and learning styles is a necessary step in building a mutually gratifying relationship.

The teaching style of any given supervisor has a variety of components, some of which promote learning and others that may impede the learning process of each individual intern. It is incumbent upon the intern to analyze his or her own characteristic learning style and present this information to the supervisor. It is not sensible to assume that the supervisor will be able to intuit what works best for you, just as an academic instructor may not always educate in a manner that is ideal for every student. Conveying what you know about yourself in this area can contribute to a supervisory experience that is effective and satisfying. Relevant and specific disclosures about teaching and learning styles pave the way for both parties to modify their styles and clarify their approaches. Whatever the outcome of the discussion, efforts of this kind have the inherent value of enhancing the intern-supervisor alliance.

Differentiating Supervision from Therapy

Another issue deserving special mention is an aspect of supervision that arises most often in counseling internships, though it is present to some degree in all human services training. This is the issue of boundaries and of what is appropriate to discuss in supervision conferences. Training in the field of human services encourages self-awareness in the areas of personal and professional development. One result of this dual emphasis is the confusion students experience as they attempt to make an appropriate selection of topics to bring to the supervision meeting. For example, a discussion of a particular client may cover reactions or difficulties in relating that can be fully understood only in light of the intern's own interpersonal and psychological makeup. Interns often voice their concerns about clearly differentiating supervision from personal therapy. They observe how some dimensions of the supervision experience seem analagous to therapy and wonder how to distinguish what is appropriate in each setting.

In general, supervision focuses on helping the intern to develop greater knowledge of self and others (as pertains to the interactions occurring in a given fieldwork site), enhance skills, obtain specific assistance when necessary and identify personal issues that may obstruct the performance of the intern's work. During this process, it is not unusual for an intern or supervisor to call attention to a personal characteristic, communication style, or specific issue that seems to require understanding or resolution. Interns may appropriately raise personal issues for discussion when doing so will benefit their learning and, ultimately, the clients they serve. Be aware of your goal in initiating such an exchange and make an effort to focus on how the particular concern is affecting your work. An experienced supervisor will help you to discover how a specific issue is becoming an interference, and will make a recommendation for personal therapy when it might be necessary or helpful. In the event that your supervisor recognizes an issue you may be struggling with, try to participate in the discussion and remain open to feedback. Approach the interaction in a nondefensive manner; it often is easier for others to identify problem areas to which we may be blind. If it seems to require further attention than the supervision context may allow, state your intention to pursue additional consultation.

VIGNETTE ONE

Kim, age 22, is an Asian student in her final year of a degree program in the human services. She is pursuing a special interest in administration. This semester she has selected as a fieldwork site a community services program that assists the homeless population by providing for basic needs such as food, temporary shelter, and transportation for medical care. Following a discussion of learning objectives in which Kim emphasizes her desire to become more knowledgeable about agency administration, she and her supervisor formulate a project idea focusing on fund-raising to extend the range of services. This is the first time Kim has been involved in a task of this kind and before long she is acutely aware of a need for help in structuring, researching, and managing several aspects of the project. Feeling unable to wait until the next scheduled supervision meeting, she approaches her supervisor (a very experienced administrator in his late forties) in passing and quickly conveys her concerns regarding the fund-raising endeavor. The supervisor does not offer any specific suggestions, nor does he propose a meeting to discuss matters further. Kim interprets this to mean that she is expected to proceed independently, and she tries to continue the project. Over the next several weeks, she immerses herself in gathering statistical information and devising a plan for particular fund-raising activities. Her supervision meetings become irregular as the middle of the semester approaches, and although she remains uncertain about her work, she decides that it is best to refrain from seeking direction or guidance. On the due date for the first draft of the project, Kim reluctantly presents her materials to her supervisor. He evaluates her progress, focusing on those items that are incomplete or disorganized. Kim is reminded of similar scenes with her father, whose positive acknowledgment has always been of great importance to her. Kim leaves this meeting disappointed and less confident about her ability to successfully complete the project. She questions whether program administration is an appropriate career choice.

QUESTIONS FOR DISCUSSION

1. What intervention, if any, might benefit Kim in the early stages of this fieldwork experience?
2. How might specific cultural factors play a part in determining whether Kim actively seeks additional direction?
3. What difficulties, if any, are suggested by the recollection of previous father-daughter experiences?
4. What aspects of Kim's interpersonal style might be contributing to her anxiety and confusion? What suggestions would you make to assist her?
5. Does Kim appear to be making any assumptions about her supervisor? about her own role in the project? Do you agree or disagree?

ANALYSIS

In this scenario, a young fieldwork student undertakes a project that involves several activities unfamiliar to her at this point in her education and professional development. Although she is interested in the nature of the assignment, she faces a dilemma when she recognizes the gaps in her own knowledge about how to best proceed. Interns often find themselves in just such a position: they are assigned an interesting and challenging task but do not have sufficient information to begin to organize and structure their work effectively. They wish to be seen as competent and confident, which for them becomes synonymous with independent and problem-free. They know they are expected to draw on their own experience and refrain from relying too much on specific direction from others. However, when difficulties arise, their reluctance to seek additional consultation can impede progress and even delay the completion of a project. Finding the optimal balance between these two points of view demands perseverance and self-awareness.

Other factors may contribute to the situation, including the cultural background of the student (and perhaps that of the supervisor), respective ages (and age differences), learning and teaching styles, and individual expectations. Decisions about seeking help can be complicated by the behaviors and attitudes learned in a particular familial or ethnic context. Gender and/or age differences can influence the interpersonal interactions that are so integral to successful collaboration. Additional information about each of these factors would add to our understanding of Kim's ability to negotiate her needs and perform her assigned tasks. We would be wise to explore how previous life experiences and intrapersonal issues might be activated in this situation and bias or distort the intern's responses.

VIGNETTE TWO

Sam is approaching the midpoint of a yearlong internship in a community clinic that provides low-cost counseling to a highly diversified client population. This is his second experience with individual and group counseling in a clinical setting; at his first fieldwork setting he was able to observe licensed professionals in both of these modalities. He also has been involved on a volunteer basis with a youth group in his neighborhood for the past five years. Recently, Sam initiated a discussion with his supervisor about the possibility of leading a group for at-risk adolescents; the idea was well received and plans were implemented to inform staff and other interns that a group of this type would soon be forming. Sam took full responsibility for developing a group format that would focus on issues related to substance abuse, sex education, communication, and problem solving. Because group membership was kept to a maximum of six individuals, Sam declined a fellow intern's offer to colead.

The first four weeks of the group went well and Sam took advantage of his supervision to explore the dynamics of the members' interactions with one another. Although he benefitted from these discussions, he already felt confident about his ability to respond effectively in the group; he credited this to the experience he had accrued working with adolescents in the youth group. In subsequent group meetings, two members began to disclose current behaviors that Sam found alarming: One member was experimenting with marijuana and cocaine; another described several incidents of unprotected sexual contact. Sam took both of these matters to his supervisor, hoping for support in his decision to respond to the disclosures only by giving advice about the possible consequences of such reckless actions. The supervisor directed Sam to explore both matters further and to gather additional information that would clarify an appropriate course of action. He pointed out that these adolescents were between 13 and 15 years of age and that there might be a need to require family involvement. Sam continued the group meetings, but was hesitant to elicit any data that might compromise his relationship with the young teens. He felt torn between the need to follow his supervisor's directives and his own desire to maintain the trust he had established with the group.

QUESTIONS FOR DISCUSSION

1. Evaluate the initial steps this intern takes following the approval of his idea for an adolescent group.
2. How would you evaluate the need for a coleader? What specific factors would you consider in your decision? What advantages or disadvantages are there in leading a group alone? in having a coleader?
3. How might this student's previous experiences with authority figures factor into his present dilemma?
4. What concerns would you have if faced with the disclosures made in this group, and how might you respond?
5. Do you think that this intern correctly assessed his readiness for this assignment, based on his previous experience with adolescents? Explain your response.
6. What options exist when interns disagree with specific directions from their supervisors?

ANALYSIS

This student certainly deserves credit for the initiative he demonstrates in developing an idea that will contribute to the scope of services provided by the clinic, and for designing his fieldwork to include areas that reflect his special interests. Interns cannot always be certain that a supervisor will sanction their idea for a specific activity or even share a similar area of interest. Interns are well advised to anticipate projects they may wish to be involved in and to discuss them early on in their fieldwork. However, if you have the opportunity to remain in an internship for a substantial number of months, you undoubtedly will develop goals that were not a part of your original agenda. Thoughtfully presenting an idea even though you have no guarantee of support or approval is a worthwhile endeavor and adds another dimension of professionalism to your overall performance.

Any new experience, such as the one depicted in this vignette, will be comprised of both rewarding and challenging moments. This student takes on the task of group leadership alone; he assesses the size of the group to be manageable without the assistance of another intern. He develops the group format independently and feels little need for input as the group begins. What aspects of personality and/or work style might be suggested by this approach? An argument could be made for collaboration on any task that is undertaken at this level of professional development, particularly when the student has had little or no previous exposure to comparable assignments. In this scenario, however, experiences in the intern's background may or may not have prepared him to manage the present situation effectively. Sam feels confident about his ability to conduct the group on his own because he has had a fair amount of experience with adolescents. He may assume that what he has learned in his years with the youth group will now generalize to his current situation, providing ample tools for him to use in his new role. This idea will have to be explored in the context of the developments in the group over time. It is certainly possible that the disclosures made by two of the members pose a situation this intern has not directly dealt with before, at least not in a role that involves the level of responsibility held by a group leader in a counseling setting. Because unexpected events can present a challenge for even the most prepared intern, it is important to examine the personal expectations of interns as they undertake new responsibilities; students and supervisors can both facilitate this process.

Finally, Sam is faced with a dilemma typical of human services settings where issues of trust and confidentiality, as well as the obligation to take reasonable actions to protect clients in dangerous situations, are a part of everyday interactions. In particular, this student faces the possibility of having to take a course of action that may jeopardize member-leader relations; this is a reasonable concern and one that demands attention. He also recognizes that the supervisor may ultimately direct him toward interventions with which he may not agree. Working through this conflict requires a willingness to present and discuss differing ideas while remaining open to direction and guidance. There is a slight indication that Sam may choose to avoid this eventuality, as suggested by his hesitation to promote further frank discussion in the group. This strategy will hinder his learning and will not enhance his future problem-solving abilities when, inevitably, a similar problem arises. Furthermore, his refusal to assume the responsibilities of his position could obstruct progress in the group.

VIGNETTE THREE

Anne is in the third month of fieldwork with a community foundation that provides services to patients with autoimmune deficiency syndrome (AIDS). The clientele served by this agency are in various stages of the disease, with the majority being confined to their homes. Others are able to come to the agency on a somewhat regular basis for needs assessments, support groups, and assistance with benefit forms. For the most part, staff members are friendly and helpful to the student interns and appreciate their substantial contributions to the overall functioning of the program. Like many nonprofit agencies, this facility constantly struggles with the intense demand for services while lacking an adequate number of staff and sufficient funds. Interns and staff members perform similar tasks and everyone is involved with direct client services, in person and by telephone.

In a weekly fieldwork seminar, Anne discusses her experiences with other students and with the seminar instructor, focusing mainly on increasing her awareness of issues related to death and dying. Generally, she assesses herself to be adapting reasonably well to a stressful environment. Assimilating information regarding the extensive system of paperwork involved has been challenging and is the aspect of the internship she least enjoys. In a recent meeting with her supervisor, she receives negative feedback regarding the quality of the paperwork she has been producing on behalf of clients. She is somewhat surprised, because no mention of deficiencies has been made prior to this discussion. The supervisor states that Anne had been given ample training and is expected to identify problem areas and take steps to improve her performance. Correctly completed paperwork, she emphasizes, is critical to obtaining necessary services in a timely manner for clients in severe need.

After the meeting, Anne seeks input from a friend who is also interning at the foundation (although he is under the supervision of a different department head). He tells her that he too was overwhelmed with the paperwork during his first month. Since then, he has managed to cultivate a level of comfort and competence owing to the patient and methodical teaching of his supervisor. Anne and her friend converse briefly and then part to continue their respective tasks. Anne is confused about how to proceed and can't help comparing her supervision experience to that of her friend, who appears to have been much better prepared. She wonders about her supervisor's evaluation of the other duties in which she has been engaged, acknowledging to herself that person-to-person contact with clients is her strength and this is certainly a more valuable skill than a talent for completing forms.

QUESTIONS FOR DISCUSSION

1. How might students cope effectively with negative feedback or critical comments from supervisors?
2. How would you advise Anne to manage her immediate personal reactions to the supervision meeting?
3. What course of action might this intern take to correct the deficiencies in her paperwork?
4. In the view of this intern, the ability to relate to clients outweighs the importance of correctly completing paperwork. What, if any, are the possible ramifications of this determination in her current fieldwork setting?

ANALYSIS

Students training in a fieldwork site such as the one depicted in this vignette face a host of difficult tasks. Working with a population that is severely or terminally ill demands patience, sensitivity, and a capacity to adapt to constant change and loss. It is also an arena in which stressful events are continuous and the student-supervisor relationship often functions as a much-needed source of support. It appears to Anne that she has made an adequate adjustment to the environment and the clientele. She actively discusses her experience in fieldwork seminars and has made the area of death and dying the focal point of her learning objectives. It certainly is not difficult to empathize with her reactions to her supervisor's critique and to the subsequent conversation with the other intern. Because it is relatively early in the internship, the response she makes to the present dilemma is likely to determine the quality of her experience in the remaining months.

In most human services settings, there is some degree of paperwork to manage, and unique methods for completing forms and general record-keeping must be mastered. Even students who are interested in the administrative aspects of human services delivery systems can find themselves feeling perplexed and overwhelmed in their attempts to integrate large amounts of new information in a relatively short period of time. In addition to coping with this task, the student in this vignette also must respond professionally to the feedback she is receiving from her supervisor. This response will be affected by her ability to understand and manage her personal reactions to criticism, and by her capacity to communicate assertively and to problem solve effectively. She might overlook the nature of the specific complaint of the supervisor by redirecting her focus to other elements in the setting (for example, comparing the supervision styles of the various department heads, gathering impressions from other interns and staff members about their experience with her supervisor, or interpreting the criticism about her paperwork skills as an indication of her supervisor's negative view of her overall performance). It will take maturity and a thoughtful processing of this experience to shape a response that addresses the concerns of both parties.

Suggested Readings

Benshoff, J. M. (1993). Developmental group supervision. *The Journal for Specialists in Group Work, 15,* 225–238.

Berger, S. S., & Buchholz, E. S. (1993). On becoming a supervisee: Preparation for learning in a supervisory relationship. *Psychotherapy, 30,* 86–92.

Bogo, M. (1993). The student/field instructor relationship: The critical factor in field education. *The Clinical Supervisor, 11,* 23–36.

Burns, C. I., & Holloway, E. L. (1989). Therapy in supervision: An unresolved issue. *The Clinical Supervisor, 7*(4), 47–60.

Hayes, R. L. (1990). Developmental group supervision. *The Journal for Specialists in Group Work, 15,* 225–238.

Heppner, P. P., & Roehlke, H. J. (1984). Differences among supervisees at different levels of training: Implications for a developmental model of supervision. *Journal of Counseling Psychology, 31,* 76–90.

Johnston, N., Rooney, R., & Reitmeir, M. A. (1991). Sharing power: Student feedback to field supervisors. In D. Schneck, B. Grossman, & U. Glassman (Eds.), *Field education in social work: Contemporary issues and trends.* Dubuque, IA: Kendall/Hunt.

Kadushin, A. (1985). *Supervision in social work* (2nd ed.). New York: Columbia University Press.

Teitlebaum, S. H. (1990). Supertransference: The role of the supervisor's blind spots. *Psychoanalytic Psychology, 7,* 243–258.

Chapter Four

Developing Skills in Critical Thinking: Legal and Ethical Issues

- Identifying Legal and Ethical Issues
- Professional Codes of Ethics
- Developing a Decision-Making Approach
- Issues Related to Professional Training
- Vignettes
- Issues Related to Helping Relationships
- Vignettes
- Issues Related to the Practice of Counseling
- Vignettes
- Suggested Readings

In many academic programs, students are given some instruction about the legal and ethical issues in human services prior to beginning the fieldwork portion of their training. Such a course may be a core requirement in a degree program. In addition, discussions of relevant legal and ethical issues are an ongoing aspect of most human services classes. Educators at the undergraduate and graduate level aim to assist students in acquiring not only a foundation of knowledge about legal and ethical matters, but also a practical approach to decision making, one they can use when faced with the many ambiguities often characterizing these legal and ethical dilemmas. In this chapter, we focus on strengthening your ability to identify and explore prevalent legal and ethical issues. We will not describe every content area at length, although you will find a list of suggested readings at the end of the chapter. Rather, we will present brief discussions followed by pertinent illustrations of actual fieldwork situations. We encourage you to imagine yourself in the role of the student in the vignettes and apply your most sophisticated critical thinking skills to each of the dilemmas presented.

Identifying Legal and Ethical Issues

Legal and ethical concerns are of primary importance in the helping professions. In recent years, increased attention to these matters has prompted a keener awareness of ethical practices on the part of consumers. The most commonly recognized issues relate to aspects of professional training, helping relationships, and the practice of counseling:

legal and ethical issues relating to professional training—scope of practice; licensure and credentialing standards; continuing education; issues related to competence, supervision, and consultation; multicultural awareness and education

legal and ethical issues relating to helping relationships—the role of values; dual relationship issues; informed consent; record-keeping; confidentiality and privilege; exceptions to confidentiality to protect client or others

legal and ethical issues relating to the practice of counseling—practice management issues; managing referrals and termination; dealing with transference and countertransference; the duty to warn in dangerous client situations; recognition of a client's right to self-determination

The issues identified here are among the most important ones facing human services workers in various types of research and direct service positions. As a student, you must acquire a working knowledge of the basic principles underlying these issues and a general sense of their historical development. Once you enter a fieldwork setting and become involved with particular kinds of helping relationships and delivery systems, make an effort to anticipate the specific legal and ethical situations that would be most likely to arise. Initiate discussions with your supervisor and other staff members about their experiences. Although it is almost inevitable that you will be faced with dilemmas you feel somewhat unprepared to manage, learning from the experience of others will give you valuable insights. Remember that all students face these issues, and supervisors expect you to seek guidance in your decision-making process.

Professional Codes of Ethics

Each of the helping professions has developed a code of ethics designed to guide the behavior and decisions of its members. Generally speaking, these codes are devised by experienced professionals with expertise in various branches of the human services. The aim is to formulate protective policies that will be mutually beneficial to both practitioner and client/patient by promoting a level of care that ensures client safety and welfare. Community agencies, hospitals, educational facilities, and other fieldwork settings may have *specific* guidelines and requirements. You should become informed about these codes and conversant in their basic principles.

Developing a Decision-Making Approach

Once you become somewhat familiar with the basic standards of ethical practice in a given area of human services, a particularly disconcerting reality begins to emerge: More often than not, the application of ethical principles is a complex and ambiguous process. In each situation, multiple factors must be considered from various perspectives. What appears to be a clear course of action is often revealed, upon closer examination, to be an imprecise strategy for arriving at a satisfying if not conclusive response. Professionals and students alike benefit from collaboration and open exchange of ideas regarding the identification of ethical issues and the most appropriate response. Like many practitioners who have been in the field a number of years, we struggle with these issues in the educational settings, training forums, and agencies in which we work, as well as in our private practice. We have come to recognize the value of developing a systematic way to think through ethical and legal dilemmas. Our decision-making approach includes consultation, critical deliberation, and a willingness to consider each particular dilemma from various perspectives. We encourage you to engage in a similar process as you consider each of the following situations, which present common ethical and legal problems in fieldwork. Use your fieldwork seminar as a resource for discussion, and leave yourself open to discovering alternate ways of understanding principles, perspectives, and creative applications of ethical practice.

Issues Related to Professional Training

VIGNETTE ONE

You are in your first semester of fieldwork at a child guidance center that provides mental health services to families and children in your community. You are a graduate student pursuing a degree in counseling and are looking forward to taking on a caseload of your own. After meeting with your supervisor, you are assigned to a family that has been seeing a staff counselor for child therapy and family meetings for the past year. Due to the therapist's upcoming maternity leave, the family has agreed to transfer to a new counselor in order to continue their work on several parenting and family issues. Although they have consented to the transfer, they are ambivalent about making the transition and have expressed doubts that anyone will be able to fill the shoes of their current counselor.

You contact the family to arrange an initial visit. In this telephone conversation the parents express doubts that this new arrangement will succeed and openly describe the feelings of loss and disappointment they are experiencing due to their previous counselor's departure. You learn at this time that they have had three counselors at this agency over the past few years, two student trainees followed by a seasoned staff member. During the first visit with the parents, you introduce yourself and begin to focus your attention on fostering a new therapeutic relationship. You assess the family's goals and begin to determine a course of continued work on specific issues. Although you have discussed with your supervisor the need to disclose your status as a trainee, you elect to spare the family what you believe might well jeopardize an already fragile alliance: the announcement that they are now working with an unlicensed, supervised student counselor at the beginning of her training.

QUESTIONS FOR DISCUSSION

1. What would you identify as the relevant legal/ethical issues in this situation?
2. Looking at the situation from the perspective of the family, what are the pos-

sible effects of the student's decision not to disclose her trainee status? from the perspective of the student?
3. Consider the issue of supervision from the point of view of the family members. What information are they entitled to receive about this process?

VIGNETTE ONE (CONTINUED)

In your next supervision meeting, you review the first family session and refer vaguely to your concerns regarding the transition from one counselor to another. Although you have mixed feelings about the decision to refrain from disclosing your training status to the clients, you are aware that your supervisor would most likely disapprove of this strategy. As the discussion continues, the supervisor asks directly how the clients responded to the knowledge that you are a trainee. You explain your hesitation to burden them with that information so soon in the transition and your wish to protect the newly developing relationship. The supervisor directs you to rectify this situation when you next meet with the family (there is no time left in the meeting to discuss the matter any further). When you inform the family that you are a trainee, they receive the information quietly and without discussion. However, your worst fear comes true when the family fails to keep any subsequent appointments.

QUESTIONS FOR DISCUSSION

1. How might a discussion have proceeded between the student and the supervisor regarding the ethical and legal ramifications of failing to disclose the student's status as a trainee?
2. Assume the role of the student in the vignette. What personal issues might be involved in the student's reluctance to present sufficient information about her status and level of training?
3. What possible factors might explain the family's decision to discontinue therapy?

ANALYSIS

Students in training and fieldwork often are painfully aware of their position in the hierarchy of their fieldwork setting. This awareness, along with the desire to be successful, to feel competent, and to be evaluated positively by supervisors and colleagues, may prompt them to shield their work from observation or critical examination. As the above vignette suggests, students can be tempted to withhold essential information from those they seek to help ostensibly to protect themselves or others from failure. This issue is an appropriate topic of discussion in fieldwork seminars and supervisory conferences.

Ethical and legal aspects of this situation include issues of competence, accurate representation of level of experience, the need for supervision, and the right of the client to informed consent. Those in training settings must cope with the frequent beginnings and endings that characterize fieldwork assignments lasting one or two semesters. Clients typically are informed about this aspect of the setting at their initial interview, but few can understand the effects of staff turnover prior to experiencing it. It is clear that the clients in the vignette have experienced less than ideal circumstances during the course of important therapeutic relationships.

You undoubtedly will be faced with some form of this dilemma as you move from one fieldwork setting to the next. Remember that informed consent can exist only when sufficient information is provided to clients. When information is presented in a straightforward, open manner, clients are reassured that you are up to the challenge of dealing with the situation and feel empowered to make a good decision on their own behalf.

Students in training, although limited as to their acquired professional experience, have a tremendous amount to offer. We encourage you to realistically appraise your abilities on an ongoing basis throughout your fieldwork months so that your confidence as a developing professional will continue to grow. In this way, you also will be more inclined to use supervision in a truly productive manner. Supervision is an aspect of your training that should be both supportive and educational. Too often, however, students who are struggling with shaky self-confidence interpret their need for supervision as a further indication of incompetence. As a student, you may fear that a client who is informed that you are working under supervision will doubt your capacity to be effective. Three important issues pertain: the right of your client to informed consent, accurate representation of your competence, and the issue of confidentiality as it relates to supervision.

The various aspects of the professional training process in the human services are best understood by those who find themselves in that process. The average consumer of helping services is largely unfamiliar with the distinctions that are made between types of educational degree programs, training and fieldwork, credentialing, and state licensing requirements. You are the primary source of this information as a client strives to make an informed decision about entering into a helping relationship. You must provide relevant details and invite questions. Keep in mind the following as your guide: What would *you* like to be advised of as a consumer in a similar circumstance?

The ethical and legal aspects of supervision relate most strongly to issues of confidentiality and privacy. (Refer to the end of this chapter for a list of readings to increase your knowledge about these concepts.) Students in counseling confront these issues most frequently, although those in the areas of education, health care, social work, and public administration also may encounter them. As a fieldwork student, you are expected to engage in open discussions about your work either in individual meetings with your supervisor or in a group format. Both of these forums present some challenges when it comes to respecting the principles of confidentiality and privacy. What information is essential to present in supervision? Should information revealed in supervision be general or specific? What becomes known in the process of documenting a client's assessment and progress, and who will have access to that file? What does the client have a right to know regarding the supervision format and the extent of disclosure that occurs therein? What are the exceptions to confidentiality and when should clients be informed of them? Discuss these common concerns in your fieldwork seminar, and try to develop your own unique approach to making ethical decisions. Reflect on the dilemma of the student in the vignette and revise the course of action to reflect your own conclusions regarding ethical and legal principles.

VIGNETTE TWO

You are an undergraduate human services student in your second semester fieldwork at an agency that provides home visits to Alzheimer's patients and their caregivers. The majority of your clients are senior citizens in a primarily Hispanic community. Prior to the past month, you were performing administrative duties at the home office; you have just recently begun additional work out in the field. Although you are excited by this new opportunity, you often feel confused by the reactions of the caregivers to your attempt to gather information about the needs and welfare of the patients. Your direct questions seem to elicit only brief responses. Whereas you had expected that the caregivers would be happy to have a helping person involved in this difficult family situation, you are aware that frequently your visits seem to cause discomfort and irritation. The time allotted for each visit is approximately an hour, but you generally find yourself concluding your visit in less than half that time because of the anxiety you feel about holding successful conversations with the clients. When making entries describing your visits in clients' files, you rely on your observations of the family and are unable to provide much in the way of specific updated material. After a few weeks of feeling frustrated and inadequate, the need for supervision becomes exceedingly clear to you.

QUESTIONS FOR DISCUSSION

1. What are the potential ethical issues facing the student in this vignette?
2. What are the major factors contributing to the student's frustration, from the point of view of the student? of the caregiver? of the patient?
3. What are the ethical responsibilities of the student in deciding how to proceed? of the supervisor? of the agency?

ANALYSIS

One of the most appealing aspects of the field of human services is the number of opportunities for exposure to a diverse spectrum of people and life experiences. Students often describe the excitement inherent in being involved with populations, cultures, and settings outside their usual frame of reference. However, students and professionals in the field must constantly orient themselves to both obvious and subtle distinctions in each new circumstance they encounter. This often involves continually learning new tasks, honing interpersonal skills, and discovering how to be most effective in performing their responsibilities. The student in the above vignette is faced with exactly this kind of challenge.

Like most students in fieldwork assignments, this student appears to be anticipating the home visits with a positive outlook. In order to make a successful transition from performing administrative tasks to undertaking the role of home visitor, it would be wise for the student and supervisor to collaborate in assessing the student's readiness to perform the functions required in community outreach work. Issues of competence are central to the idea of ethical practice. The following interpersonal skills will prove important: the ability to establish rapport, to communicate clearly in a nonthreatening manner, to gather information through focused questioning, and to be alert to nonverbal cues. Learning about common experiences of Alzheimer's patients and their caregivers will allow the student to develop questions that are relevant and productive and that demonstrate enough familiarity with the subject to reassure clients during initial meetings.

Even the most articulate and sensitive communicators will find their efforts hampered when cultural factors are given inadequate consideration. The cultural components of age, ethnicity, and client population are the prominent factors in the above vignette. Workers in the human services must also investigate family mores, language, and socioeconomic status in order to more fully understand how to set the stage for a successful helping relationship. The potential for a cooperative working partnership is increased when helpers demonstrate respect for and awareness of differences. Be conscious of your own tendency to assume that others share your frame of reference.

If you were the student in the above situation, what steps might you take to adapt yourself to the cultural milieu described? What aspects of the situation might you have a tendency to overlook? Which would present the most difficulty for you?

Issues Related to Helping Relationships

VIGNETTE THREE

You are a social work student doing fieldwork at a facility that provides emergency shelter to children temporarily placed in protective custody. Most of the children are from home situations in which they have been physically and/or sexually abused. You are one of several student interns working on a unit comprised of fifteen adolescents. There is always a regular staff member present to oversee and manage the daily routine. Weekly group supervision is provided to all interns by the director of the facility. While on duty one day, you overhear a female adolescent describing a situation in which she was molested by a relative who currently resides in the family home. She is disclosing the information to another intern with whom you have frequently worked and is insisting that the matter be held in strictest confidence. Because you and the other intern work alongside each other for the remainder of the shift, you become aware that the intern fails to document the adolescent's disclosure. Having read all the files on current residents, you know that the information about the sexual molestation has not been previously related.

QUESTIONS FOR DISCUSSION

1. What legal and/or ethical issues must the student in this vignette consider?
2. Apply a decision-making approach to these legal and ethical issues in order to determine possible courses of action. What resources might you consult?
3. What is the responsibility of an intern who unintentionally becomes a party to this type of confidential disclosure?
4. What relevant ethical principles might guide your decision-making process?

VIGNETTE THREE (CONTINUED)

When the next group supervision meeting occurs, all the interns are participating in a discussion about significant events that have taken place on the unit over the past week. You find it hard to think about anything other than the adolescent and her disturbing disclosure. You are aware that the residents in this facility have an average stay of only a few weeks. It has been your experience that discharges sometimes occur quite suddenly owing to rapid developments in family, court, or social service matters. During the course of the supervision meeting, you listen intently and hope that the intern who received the disclosure will raise the topic for discussion. When this does not occur, you feel confused about the role you should take in resolving the problem. You are concerned about the welfare of the adolescent and want to be helpful to her. You also are mindful that in a previous fieldwork setting you had some difficulty maintaining harmonious relationships with other interns and supervisors; you have no wish to damage the good rapport you have established at this agency.

QUESTIONS FOR DISCUSSION

1. Strong working relationships are vital for ensuring optimal functioning in human services environments. If you were the student in the above dilemma, how might you proceed?
2. What factors would you consider in making a satisfactory decision?
3. How might a supervisor be most helpful to a student in this circumstance?
4. What personal characteristics might enhance or inhibit your ability to move toward a resolution of this problem?

ANALYSIS

We certainly can empathize with this student's concerns about being of assistance to the adolescent resident without sacrificing the rapport she has established with other interns. Students in our seminars frequently express frustration at the behavior of interns, staff members, administrators, or supervisors who act in ways that seem to circumvent ethical responsibilities. We all tend to feel anxious when circumstances require us to take a position that may place us at odds with others, especially when we are not yet comfortable in our professional personas. We are committed to the helping relationships we've established and naturally shy away from actions that might alienate those we intend to serve. Situations such as the one described in this vignette place us squarely in the middle of this quandary. When a person's safety and welfare are at stake, we recognize the necessity of taking protective measures; still, we feel hesitant to elicit the client's frequently unfavorable responses. Anger, resistance, betrayal of trust, and a damaged helping alliance are but a few of the ramifications of breaches of confidence. Justifying our actions by pointing out the legal and ethical mandates does little to reassure clients who must now adjust to whatever consequences occur as a result of the disclosure.

Explore your values and consider how they might affect a situation similar to the one described in the vignette. How might the issues of dual relationships contribute to the potential for a conflict or misunderstanding to develop when a breach of confidence occurs? Frequent discussions in fieldwork seminars may help to clarify your own perspective about these matters. Remember, developing a systematic approach to critical thinking and evaluation is of great importance in successfully dealing with legal and ethical issues as they arise in practical situations. Precisely correct or singular responses to these dilemmas are uncommon. As you gain experience in your fieldwork activities, you will develop increased confidence in your capacity to make appropriate, creative decisions. You also can benefit from the support and guidance that adequate supervision can provide as you navigate this delicate territory.

VIGNETTE FOUR

You have approached a local agency to inquire about a fieldwork placement that would allow you to acquire crisis intervention skills by working with pregnant teenagers. The director informs you that the agency is affiliated with a nearby Catholic church and that the majority of volunteers are members of this church. However, because there is a great need for student interns with some counseling background, individuals from various denominations are welcome. You are somewhat concerned about the approach the agency recommends for counseling the teenage clientele, but you feel reassured when the director says the center provides information about all available options to the young women who come for services.

In your first month you learn about general agency procedures and participate in a crisis intervention training program that emphasizes effective communication and education about pregnancy and parenthood in adolescence. Shortly after you begin having individual meetings with a small caseload of teenagers, a new director comes on board. At an all-staff meeting, the new director introduces himself and speaks at length about the philosophy he expects to see reflected in the counseling services provided at the facility. You become aware that this individual is strictly aligned with Catholic doctrine on matters concerning reproduction; this stance will require that you counsel your teenage clients according to a particular agenda. Other volunteers and interns do not express any feelings of concern regarding the new director's policy, so you are reluctant to initiate any discussion about the change. Your own beliefs about abortion are unclear, and this increases your anxiety about future contacts with clients who may wish to consider this option. You are more than a third of the way through your commitment to this fieldwork placement and cannot entertain the idea of moving to a new setting.

QUESTIONS FOR DISCUSSION

1. What ethical issue(s) can you identify in the above vignette?
2. How might the helping relationships be affected if this student remains conflicted about her own attitudes and approaches?
3. If you were the student in this situation, what responsibilities would you have to your adolescent clientele? to the agency? to yourself?
4. What possible courses of action, if any, would you consider?

ANALYSIS

The role of personal values in human services work has deservedly received much attention in the last two decades. As we encounter various populations and problem situations, we find it increasingly important to be alert to the many obvious and subtle ways that our values influence our behavior, perceptions, and decision making. At times, our values may be so elusive that we become fully aware of them only upon a retrospective analysis of our approach to a particular predicament. A major task of professional development—indeed, of growth into full adulthood—is to discover and clarify personal values. All of us evolve in the context of families and larger societal systems wherein we incorporate an entire spectrum of value-laden beliefs, perceptions, and attitudes. Part of achieving our own, fully autonomous identity involves sifting through the values we accepted as children to determine what we as responsible adults truly believe and support. It is an ongoing process as we gain more life experience and deepen ourselves emotionally, intellectually, and spiritually. Not surprisingly, human services workers often are faced with situations that prompt an immediate and active struggle to elucidate personal values and understand the values of others.

Imagine yourself in the position of the student in the above vignette (or in a similar circumstance involving conflict of values). In your early interview with the first agency director, your initial anxiety about the potential for a conflict of values is alleviated, and you begin counseling the adolescent clientele in the sensitive area of reproductive choices. When the directorship changes hands you experience a resurgence of that initial anxiety, made worse by the lack of opportunity to interact with colleagues about the policy shift. You have ethical concerns about the lack of clarity you detect in your own value system and about what appears to be a mandate to impose the beliefs of the agency and its director on a clientele who are confused and in crisis. To complicate things further, practical considerations about accumulating sufficient hours to complete your fieldwork requirements eliminate the option of concluding your fieldwork elsewhere.

Personal ethical and moral beliefs can often conflict with those of colleagues, clients, agencies, and policy makers. What is individually felt to be a correct course of action can become a source of dissension and division when a group or even just two people attempt to work together harmoniously. The intensity of emotion that surrounds controversial issues points up the investment we all have in protecting our values from attack. We may be extremely resistant to hearing different opinions or participating openly and flexibly in a discussion in which others are voicing alternate views. There may be a tendency to judge the beliefs of others as incorrect or morally lacking without allowing for a full hearing. Again, we encourage you to use supervision meetings and seminar classes as forums in which to discuss issues and values that arise within the context of your own fieldwork experiences. If you participate in these discussions with the intention to maintain an open mind, you will discover the sticking points that for you complicate the issues. Try to notice when your attention drifts or when you feel uninterested in listening to another's perspective. Become aware of anger, tension, or a tendency to dismiss a point of view that is contrary to your own. Know that if you let go of a comfortable attachment to a previous conviction, you may be left with a discomforting sense of ambiguity. The potential displacement of values once held inviolate is a disconcerting aspect of self-examination. It takes a great deal of courage to suspend adherence to a particular perspective while reviewing its origin and questioning our commitment. Remember that colleagues, instructors, and other students face a similar struggle and will likely be empathetic and supportive.

Issues Related to the Practice of Counseling

Vignette Five

You are an undergraduate human services student who is experiencing uncertainty about decisions regarding future graduate school studies. You are interested in both counseling and social work and are hoping that your fieldwork will provide some clarity and sense of direction as you near graduation. For your final two semesters, you opted for an internship at a social services agency that focuses primarily on prevention programs for adolescents. This seemed to be the ideal choice, because it will afford opportunities to participate in paraprofessional counseling and at the same time learn the nuts and bolts of case management. You will be expected to identify the needs of the adolescents assigned to you and provide assistance with referrals to suitable community resources.

By the beginning of your second semester at the fieldwork site, you have established a small caseload of at-risk adolescents. You meet with each teenager one time each week. It was a surprise to discover that developing rapport with this population came much more easily than you had anticipated; in fact, you had always thought working with this age group would be a difficult and unsettling experience owing to the mixed feelings you carry about this period in your own development. You particularly look forward to your weekly meetings with a 15-year-old female client whose complex family background and economic problems present substantial challenges. Your role in this case has been to provide academic guidance, encouragement, and emotional support to this teenager who is at risk for such difficulties as early pregnancy, gang association, and substance abuse.

Lately, you have been preoccupied with this client's complicated situation; you find yourself distracted from other activities as you attempt to devise creative problem-solving strategies on her behalf. Even when you are talking with other clients or attending your classes at school, you're thinking about this client. You determine that she needs more frequent contact, and you make additional time in your schedule to accommodate extra meetings. Although it is against agency policy, you give the client your home telephone number and encourage her to call whenever she feels the need. During weekly supervision meetings, you discuss the client's circumstances and your efforts to assist her. When your supervisor suggests that you may be too involved, you feel certain that no one outside the situation can truly appreciate the urgency of this adolescent's predicament.

Questions for Discussion

1. What indications are there that the student in this vignette may be struggling with an ethical issue? What ethical issue(s) do you identify?

2. Do you think that the student's reactions to this adolescent will benefit the client, or will they hinder potential progress? Explain your response.
3. The supervisor presents the student with an alternate view of the dynamics occurring in the therapeutic relationship. Because she is the only member of the agency in direct contact with the client, the student seems inclined to give the supervisor's input little consideration. What suggestions would you offer the student as to how she might benefit from supervision in this situation?
4. Keeping the client's best interests primary, would you suggest that the student refer the client to another worker at the agency? Why or why not?

VIGNETTE FIVE (CONTINUED)

Over the next several weeks, the adolescent begins to reveal a strong desire to leave home. She describes being very unhappy that her parents are unable to provide many of the material things that some of her friends at school possess. She devises a plan to reside with an older friend in a nearby city and to work at whatever job she can find there. Although she knows her family will not approve of her friend or her plans, she is certain that a better life awaits her if she only follows her heart. You are alarmed by the determination she exhibits, and you become even more worried when she is unresponsive to any suggestions you make that are at odds with her desires. In a final attempt to dissuade her, you begin to give her small amounts of money and a few of the material items you think would make her more comfortable among her classmates. Although you're relieved that this effort does seem to be delaying the adolescent from acting on her impulses, you also are experiencing a fair amount of confusion about your behavior. You voice the dilemma in your weekly fieldwork seminar and ask for feedback. In the process of discussion, you find yourself comparing the client to your younger sister. You briefly describe how, at the age of 15, she ran away from home and encountered many serious difficulties. Although you were only a teenager yourself at the time, you have always felt frustrated and guilty about not being able to help her.

QUESTIONS FOR DISCUSSION

1. Assume the role of a classmate in this student's fieldwork seminar. Having heard the facts, what feedback might you offer?
2. Imagine that the student were to continue to "help" the adolescent in the manner described. What circumstances might potentially develop?
3. Now that you are aware of the student's relationship to a sibling, and of important events associated with that relationship, how do you think past experiences influenced the student's ideas about working with an adolescent population?
4. What alternate suggestions might you make for intervening with this client?
5. What fears might you have in discussing a situation comparable to this with your supervisor or your peers?
6. What factors would you consider when deciding whether to ask for supervision?

ANALYSIS

The student in this vignette experiences a complex process known in the counseling profession as countertransference. *Countertransference* is the component of therapeutic interaction that describes the reactions of counselors to their clients. Some psychoanalytic and self-psychological approaches to counseling and psychotherapy define countertransference broadly and view it as encompassing all reactions on the part of the counselor. Others speak to a more limited scope of reactions, those that reflect only the counselor's *projections*. Our everyday interpersonal interactions tend to be significantly influenced by the important people and events in our past. Think of how many times you have responded to someone as though they possessed qualities or had expectations that are typical of your mother, father, sibling, or other significant individual. When this process occurs on the part of the client and is directed toward the counselor, it is known as *transference*. In the therapeutic setting, clarification of the transference process provides information about some of the central themes of the client's difficulties, especially in the area of relationships.

In a similar manner, an understanding of the *counter*transference experienced by the counselor in response to a client can be of tremendous benefit. Common reactions include a need for approval, identification with the client, sexual and/or romantic feelings toward the client, a tendency to refrain from confrontation, or a compelling need to rescue the client. These are but a few of the general areas into which our many idiosyncratic reactions can be categorized. These reactions become the subject of ethical consideration when they are intense, persistent, and compelling, because their presence compromises therapeutic efforts that are otherwise well founded. The onset of feelings of countertransference can be quite subtle. In this vignette, the student moves through a progression that begins when she contemplates the idea of working with adolescents. Next, the student experiences an appreciation for the rapport that is established more easily than expected. She then develops particular empathy for one female client, as well as a growing concern about her welfare. Her level of investment in the client increases as she makes her personal telephone number available and arranges for more frequent contacts. The student becomes aware of an evolving preoccupation with this client and her problems; she knows she does not expend the same amount of mental energy and problem-solving efforts with the other adolescents.

Reading about this student's predicament, you probably find it quite easy to discern the thoughts and behaviors that could indicate the presence of countertransference. However, to the person experiencing them, the subtle complexities of a therapeutic interaction can be confusing and even painful. The intensity of the emotions involved make the situation even more troublesome. When feelings are particularly intense, we may become convinced that we are assessing the circumstances accurately. Note how the student in the vignette is tempted to rely on her own feelings and responses and views the supervisor's input as probably less accurate.

We can commend the student who brings this dilemma into the open and invites feedback and discussion. The willingness to examine and explore our reactions to clients serves to protect the therapeutic interaction from harmful contamination or derailment. Progress can hardly thrive when we become prisoners of our own psychological makeup and fail to see the client as a separate and unique individual. It is unlikely that we will be able to respect a client's right to self-determination (another ethical issue pertaining to counseling) if our vision is clouded by the shadow of persons or events in our own history. How can we assist clients to discover and achieve their goals if we are subconsciously imposing our own agenda on them? We must strive to deepen our own self-awareness in order to avoid unknowingly using our clients to satisfy our own needs.

VIGNETTE SIX

You are completing the final months of a fieldwork assignment in an outpatient chemical dependency program. The program is designed to provide continuing care for adults who have completed inpatient detox and treatment at the adjacent hospital facility. For the past few months you have been coleading a men's group for adult alcoholics, developing and enhancing your skills in group facilitation. You have witnessed a range of events in the group, including tense conflicts among members and expressions of true empathy and genuine efforts to support one another. When you arrive at the facility on one particular evening, you are informed that your coleader will not be present for the group meeting due to a family emergency. Because you would prefer not to lead the group alone, you try to locate a staff member who might be available to assist you. Your efforts prove futile, so you quietly acknowledge your anxiety and begin to gather your thoughts in preparation for the start of the meeting.

The group arrives and settles in for the evening's session; all twelve members are in attendance. The major theme of the discussion is anger—how to express it appropriately and recognize it early enough to prevent exacerbating a conflict. One of the more outspoken members confronts another member regarding the "threatening tone of voice" that he used in an interaction that occurred during the previous weekly meeting. The tension grows quickly between the two men and your attempts to intervene go unnoticed. One of the men discloses that he is not sober, having relapsed earlier that same day. He rises to leave the meeting, moving unsteadily toward the exit. From the parking lot, you hear him making threats toward his estranged wife, whom he blames for his addiction. With little time remaining, you facilitate a group process focusing on the conflict and the departure of the inebriated and agitated member. One of the quieter members of the group takes the last five minutes of the session to describe the depression and hopelessness he has been experiencing for some time now. In an effort to protect his family from any further hardship, he refuses to let them know that he has lost yet another job. He says he is not sure he can go on.

QUESTIONS FOR DISCUSSION

1. What factors might the student have considered in making the decision to lead the group alone? What ethical issues need to be taken into account in arriving at this decision?
2. Identify the ethical issues that begin to arise as the conflict within the group escalates.

3. Assume the role of the student. What steps would you take (if any) to respond to the threat made by the departing group member?
4. The intoxicated member who left the meeting may pose a risk as a drunk driver. What, if any, are your legal or ethical mandates?
5. How might you assess the circumstances involving the group member who has lost his job and feels despondent? Are there legal and ethical issues you must address?
6. In the event that you initiated a hospitalization for suicidal intentions, how would you deal with the issue of confidentiality? What would you disclose to family members?
7. What role does confidentiality play when disclosures occur in a group setting? Do group members have the same obligation to respect confidentiality as group leaders?

ANALYSIS

It is not uncommon for students and professionals in the counseling profession to encounter crisis situations like the ones depicted in the preceding vignette. Clients who pose a danger to themselves or others present us with great challenges and test our capacity to tolerate intense emotional states and maintain some measure of calm and clarity. Legal precedents and ethical mandates often are created in response to tragic circumstances involving clients and their victims. Counselors and therapists are charged with the responsibility of making informed decisions in cases involving the safety and welfare of their clients. Their job is to ensure their clients' protection and that of others who might become the objects of a client's aggressive actions.

Managing your responses to conflict and anger effectively is a skill that comes from extensive experience and training. Human services professionals should become aware of their style of dealing with conflict (something that is, in part, the result of what they learned in their family). Choices can then be made and new responses learned and practiced, so that they can be used when the need arises.

This student in the vignette decides to facilitate the group alone. Perhaps he feels familiar with the role of leadership by this time in his internship. He also may feel a sense of responsibility toward members who, due to the late hour, are already on their way to the facility. Issues of competency, scope of practice and training, agency policies regarding students facilitating groups alone, and the specifics of insurance billing all must be given careful attention when decisions such as these are made. A brief consultation with a supervisor or staff member can be very helpful when you unexpectedly find yourself faced with a similar dilemma.

In the course of the group meeting, what may have begun as a productive confrontation quickly escalates into an angry stalemate that results in the premature departure of a group member. As is typical of crisis situations, the student here gradually becomes aware of information that leads him to question the client's ability to manage his anger. Because he must respond quickly, the student may feel overwhelmed by the need to prioritize several important issues that need immediate attention. One thing he must attend to is the emotional state of the remaining group members as the meeting draws to a close. How might *you* decide to best proceed in a moment like this? Do you agree with the student's focus on facilitating a brief group process about the events that just occurred?

The final moments of the session present the student with yet another complex development, as the last group member to make a verbal disclosure describes struggling with an uncertain level of depression. After the student identifies the relevant legal and ethical implications of the client's disclosures, some course of action will need to be taken to address matters of client safety. Concerns about confidentiality are foremost; note that this client has specifically stated his wish that family members not know the extent of his current predicament.

Striking the appropriate balance between giving too much or too little information to emergency personnel or hospital admitting staff is a difficult task. Making the same call with members of the client's family can be even more formidable since the emotional reactions of families in crisis place great demands on us as helping professionals. It takes considerable practice and experience to cultivate an approach that is founded on ethical principles and critical reasoning and that provides empathy for the client and all others who are affected by elements of the crisis. In the heat of the moment, you may be confronted by a range of competing viewpoints that could lead to unnecessarily compromising your client's confidentiality. Although we may feel some relief at the prospect of sharing the gravity of a crisis with others who wish to be involved, our professional roles sometimes require that we retain confidential disclosures no matter how upsetting they may be. It is our legal and ethical responsibility to protect the privacy of our client to the greatest extent possible, which usually means providing the least amount of confidential information needed to keep the client safe.

In the final analysis, ethical and legal dilemmas in the human services figure prominently in our work. They are challenging for novices and seasoned professionals alike; the preceding scenarios illustrate the complexity of such dilemmas. It is not our intention to suggest that any single strategy is the correct one to embrace when making ethical decisions. For that reason, we have not drawn too much attention to how *we* might choose to respond to any of the specific situations presented in this chapter. Rather, we have attempted to pose pertinent questions and encourage in-depth discussion, which we hope will help you to become increasingly alert to the legal and ethical elements in your work.

Suggested Readings

Arthur, G., & Swanson, C. (1993). *Confidentiality and privileged communication*. Alexandria, VA: American Counseling Association.

Bongar, B. (1991). *The suicidal patient: Clinical and legal standards of care*. Washington, D.C.: American Psychological Association.

Corey, G., Corey, M., & Callanan, P. (1993). *Issues and ethics in the helping professions* (4th ed.). Pacific Grove, CA: Brooks/Cole.

Edelwich, J., & Brodsky, A. (1991). *Sexual dilemmas for the helping professional*. New York: Brunner/Mazel.

Herlihy, B., & Golden, L. (1990). *Ethical standards casebook* (4th ed.). Alexandria, VA: American Counseling Association.

Huey, W. C., & Remley, T. P., Jr. (1989). *Ethical issues in school counseling.* Alexandria, VA: American Association for Counseling and Development.

Linzer, N. (1990). Ethics and human service practice. *Human Service Education, 10*(1), 15–22.
MacNair, R. (1992). Ethical dilemmas of child abuse reporting: Implication for mental health counselors. *Journal of Mental Health Counseling, 14*(2), 127–136.
Natterson, J. (1991). *Beyond countertransference*. New York: Jason Aronson.
Stadler, H. A. (1986). Making hard choices: Clarifying controversial ethical issues. *Counseling and Human Development, 19*(1), 1–10.
Zakutansky, T. J., & Sirles, E. A. (1993). Ethical and legal issues in field education: Shared responsibility and risk. *Journal of Social Work Education, 29*, 338–347.

Chapter Five

Encountering Cultural Differences

- Diversity in the Work Setting
- The Multifaceted Nature of Cultural Differences
- Vignettes
- Suggested Readings

Diversity in the Work Setting

It is axiomatic that as human services workers you will encounter a variety of individuals who are different from you. The range of differences and your reactions to them is a complex equation dependent on many factors such as the setting of your internship and your own individualized training, personal history, and pattern of experience. In this chapter we will discuss some of the many variables that contribute to diversity in the work setting. It is our experience that a willingness to explore and learn about the individuals with whom you will interact, coupled with a desire to understand your own idiosyncratic reactions, is essential to success as a human services professional. In addition, you must be open to the idea of learning from your clients. A nondefensive posture and the realization that you will make errors in interpretation, just as your clients will, is necessary to survival and growth over time. Successful workers attempt to learn about their client populations in as many ways as possible in order to improve understanding and efficacy regardless of the setting or type of work.

The Multifaceted Nature of Cultural Differences

Consider what comes to mind when the word *culture* is mentioned. Most of us begin to imagine various aspects of ethnicity and familial characteristics. However, ethnicity is only one aspect of what we refer to when we talk about human services professionals dealing with cultural differences. Think for a moment about all the ways that an individual might differ from yourself. Initially you might identify differences in gender, religion, race, and so on. These are significant and meaningful descriptors; however, they do not paint a comprehensive portrait of you as a person. Other, less apparent elements are important, too, such as socioeconomic status, geographic influences, and ideological affiliations. Of course, no one person is ever fully described by any list of characteristics; accurate understanding of people requires a holistic approach. Variation among human beings contributes significantly to the satisfaction we gain as members of the field of human services. We must strike a balance between the ability to compare and conceptualize and the need to remain open in our thinking in order to avoid generalizing about our clients.

All workers need to develop a sensitivity to their own value-based positions and perspectives. One reason is that the values that we have incorporated over a lifetime always color our perceptions of others. Consider some of your own assumptions about persons who belong to various groups. Most likely, you hold some beliefs that are an outgrowth of your own cultural background rather than the result of empirical review. This is the rule rather than the exception, and it has a number of implications. To begin, one worker's approach or methods may not be compatible with the frames of reference of the clients served. For example, one group may emphasize the value of verbal expression, another may not. Nonverbal behavior has no universal connotations, nor do such concepts as "normal" or "healthy." When we are insistent on viewing all others from the perspective of our own experience, we are vulnerable to bias or ethnocentrism and ultimately to failure with our clients.

The values and belief systems of clients are as varied as one can imagine. Without question, there will be occasions when each and every human services worker is challenged by behaviors and attitudes that he or she finds disagreeable or even repugnant. For those interns who enter the world of hospitals or clinics, a wide range of behavior and emotion becomes evident. The actions of individuals who are drug dependent, terminally ill, severely depressed, or overwhelmed by conditions of abuse or poverty are not likely to seem familiar or comfortable to the uninitiated. For a student who has known the security and support of a middle-class upbringing, the role of caregiver to a family that finds itself without a home may be awkward or disconcerting.

All interns must begin to anticipate how they will react and respond to those individuals who provoke strong reactions. No matter how sophisticated our education or training, professionals and beginners all respond as human beings first and foremost. Professionals must make a continuing effort to remain honest and reflective about their responses to others. Remaining intellectually curious, open to new ideas, and flexible in one's thinking are all important to success and survival; rigidity and close-mindedness are counterproductive when one is trying to understand the behavior of others.

During the course of our academic years we are exposed to populations, systems, and groups of all sizes. Family structure is an integral component of an individual's upbringing. Social group membership is a core concept in understanding adolescent development. Environments such as the workplace or school are critical to the successful functioning of any person. From micro to macro levels, our observation of human groups or systems assists us in understanding the behavior of the individuals and populations we have as clients. Although the social sciences are influenced and directed by the hard sciences, those who are interested in human behavior recognize that science is not the exclusive province for understanding individuals, families, groups, societies, and institutions. The arts contribute much to our understanding of meaning, as does organized religion. Try not to consider one single source of information exclusively in your efforts to grasp cultural differences and become a skilled professional. Remember, too, that no one description or theory is infallible, just as no practitioner is immune from misunderstanding.

Religious Perspectives

The religious beliefs of clients will typically vary, reflecting the overall diversity inherent in the community. Without question, the religious and spiritual beliefs of your clients often will have a profound effect on their behavior and may occupy a central position of importance in their lives. Entire neighborhoods and communities may be organized around the activity of the local church. Take a moment to reflect on the role of spirituality and/or religion in your own life. Consider how your own religious beliefs and practices are an outgrowth of (or a reaction to) the experiences in your family of origin. As a developing child, what messages were transmitted in the context of your family about the meaning of family life, personal values, ethics, the value of money, relationships, sexuality, love, marriage, and work? Obviously, if we were to overlook religious experience, we might misunderstand the basic tenets of the family with whom we are interacting. Bear in mind that clients may or may not assist in our understanding of their beliefs. Many times we must ask to be educated about our clients' lives, including their religious beliefs. Quite often, individuals or families will view our sincere interest as a sign that we value and respect their individual belief systems. The worker who overlooks or, worse yet, minimizes the importance of religious beliefs risks alienating or offending a client.

Because an individual's religious training often occurs during critical developmental periods and in emotionally charged contexts, we must respect the power and prominence of this experience. The worker who expects to alter longstanding beliefs that are grounded in religious training is not only naive but also quite possibly unaware of his or her own desire to impose personal beliefs on another. Teachers, counselors, health professionals, and others are constantly in a position to consider the beliefs and actions of their clients. We must ask ourselves questions such as: Why am I upset or bothered by a particular action? How do I justify my intent to modify that behavior? Is the client/student/patient/group/community interested in the nature of the change that I propose, or am I guided in my efforts by personal beliefs or emotional reactions to the clients? How is my own religious training and spiritual philosophy influencing my views?

As a human services worker you must consider the nature of the role that you occupy in relation to your clients. Teachers, managers, nurses, child development specialists, counselors, and others all occupy positions of authority. As a consequence, their viewpoints and the manner in which they express them tend to have considerable influence compared to the opinions expressed by persons outside such roles. In general, clients anticipate receiving some direction and guidance in problem solving or task management. Human services workers should be aware of how clients view them. For example, clients tend to believe that they will be judged or evaluated by the professional, so they may strive to satisfy the perceived wishes of the evaluator. Alternately, the client may ask the worker directly for advice on how to act or what to believe. Without question, it is necessary for workers to monitor their impact on individual clients and to consider whether any boundaries have been crossed. This is not to suggest that all clients will be swayed by any suggestion made by the worker. Yet some clients will be strikingly vulnerable to the influence of another, particularly in a time of pronounced need. Remember that influence or direction need not be delivered forcefully in order to sustain an impact. The subtle suggestion, verbal nuance or inflection, facial expression, or the change in volume of one's speech may impress clients more than any direct attempt to influence. Imagine an anxious client who shares an aspect of their religious beliefs that you find to be contradictory to your own. What message do you imagine that you might convey (intentionally or unintentionally) if that particular belief is diametrically opposed to your own? Would you be tempted to refer the client to another worker? If so, you need to explore your own values and capacity for tolerance. The likelihood of encountering individuals with religious beliefs that are considerably different from your own is not remote by any means.

Consider the following questions in light of your religious beliefs and background:

1. How open are you, in general, to the ideas put forth by religions other than your own?
2. To what degree do your own religious and spiritual beliefs guide your daily behavior?
3. How should a worker respond when asked about his or her own religious beliefs?
4. Do you experience conflict between any of the ideas or concepts taught in your professional training and the tenets of your religion? If yes, please elaborate.
5. Are your perceptions of a client colored by that person's religious beliefs?
6. When, if ever, is it appropriate for a worker to challenge a client's religious or spiritual belief system?
7. How might a worker respond to a client who attempts to convert the worker to his or her own belief system?
8. When, if ever, is it appropriate to transfer a client to a worker who shares the same religious background?

Socioeconomic Perspectives

The individuals who seek services provided by human services workers will reflect the socioeconomic status of the surrounding community. Of course, some settings focus their services on particular subgroups in the area (the elderly, children, low-income families, and so on). In any event, workers must be familiar with the stressors and concerns manifested by the clients commonly seen in the catchment area. We do not advocate forming rigid ideas about any population based on socioeconomic status. However, it would be an error not to learn as much as possible about the distinctive concerns that accompany life in a defined group. For example, families that must cope on a daily basis with income that falls at or below the poverty level will experience different challenges than, say, an affluent family. Any worker who does not realize that difference and attempt to become educated about the real impact of those factors will find it difficult to establish credibility with a client. Having made that point, we urge you to maintain a flexible posture and remain willing to discard preconceptions that simply do not fit. Individual human beings vary in infinite ways. The experienced worker learns to be open to variation and surprise about all individuals and groups.

Most of us have developed a learned response to persons who are noticeably different from ourselves socioeconomically. If you don't think this is true, consider your reaction (including your internal dialogue) when you encounter a homeless person begging for change or, alternately, the sight of a sparkling Rolls Royce pulling up to the curb. Although these may represent extremes, the point is that we all tend to have some reaction to individuals relative to their financial well-being. Our culture emphasizes the accumulation of wealth and the pursuit of financial security. Human services workers will find themselves interacting with members of various socioeconomic levels and should prepare themselves for this eventuality.

It's obvious that we respond to our clients' attributes and aspirations; however, we also must consider their perceptions of us. A low-income family member who struggles to provide food on a daily basis may feel some resentment toward the worker who appears to have a strikingly different lifestyle. Similarly, an affluent couple seeking counseling may be skeptical initially about the intern who wears blue jeans to work and drives a battered pickup truck. Are any of these initial reactions appropriate? Perhaps not, but they may exist just the same and must be consciously dealt with by the worker (as must any other salient factor that affects the worker-client relationship). How and what we communicate to our clients is determined by our overall presentation, including our projection of socioeconomic status. Professionals who learn to become extremely perceptive of their clients sometimes forget that they are closely scrutinized by clients as well. As always, it is a reciprocal relationship in which actions and reactions are continuous and complex.

Although human services workers are not expected to be similar to all of the varied populations they encounter, they are obliged to remain sensitive to particular differences that may evoke strong associations for either party. Indications of one's economic well-being (or lack thereof) are usually obvious and unavoidable. Most of us respond to these symbols consciously or otherwise. It is unlikely that one could be reared in our society without some awareness of and reaction to class differences, regardless of our efforts to remain unbiased.

Here are some questions to help you reflect on your own values and acquired perceptual tendencies:

1. Describe your own socioeconomic status. Be specific as to those aspects of your self or your lifestyle that you identify as descriptors. How did you acquire the status you identified?
2. Identify the critical features that you believe distinguish you from those who occupy socioeconomic levels directly above and below yourself.
3. Try to imagine how clients of higher or lower socioeconomic status might perceive you. Do you anticipate any distortions or misperceptions? If so, state why they may occur.

4. Attempt honestly to describe any acquired views that you have regarding members of higher socioeconomic status and lower socioeconomic status. Are any of these views reflexive? Where did they originate?
5. Would you have any difficulty relating to a worker to whom you turned for help if he or she appeared to be of different socioeconomic status? If so, why? Would you say anything about it to the person?

Gender-Based Perspectives

There is little doubt that one's perceptions, interactions, and communications are influenced by gender. Most of us can call to mind any number of instances where we experienced or witnessed notable distinctions in the ways men and women were viewed or treated. Although an in-depth study of this topic is of course impractical and is beyond the scope of this text, we can engage in some meaningful discussion. We encourage you to start a dialogue with colleagues and peers and promote the sharing of information.

Although traditional male and female roles have evolved over time, many societal distinctions are still made between men and women. What we may consider to be male or female in character extends well beyond biological traits. Societal training occurs throughout our lifetime, and much of our role-influenced behavior is also gender-related. To some degree, gender-specific behavior is taught within our familial contexts and is shaped by ethnicity. Although there is no absolute measure of "male" or "female" (outside of biological distinctions, which of course include some aspects of female and male for each of us), we all have some internalized concepts of what constitutes femininity or masculinity. Advertising firms have long understood that products can be effectively marketed to men or women based upon collective expectations about gender-defined preferences and buying behavior.

Marital therapists commonly witness striking differences in the communication styles that exist for women and men. Although there are never any absolute rules regarding the behavior of human beings of either sex, striking patterns do exist. For example, men often view relationships in different terms than their female partners. Many times men are observed to be more focused on action and on the concrete steps involved in solving a particular problem; the female partner may emphasize interaction and process, and focus more on the emotional components that are present. Men and women describe the verbal and nonverbal communication process differently, using different language and reflecting divergent viewpoints. It is not uncommon to hear a man and woman define their relationship or specific problem areas in differing ways due in no small part to their social learning, which encouraged specific skills and deemphasized others. The combined strength of this interweaving of skills also may be the source of many conflicts. That differences will exist is inevitable and not necessarily bad.

Human services workers will observe many decisions that appear to be gender-based. On such an occasion, workers need to analyze and attempt to understand the actions of those involved in order to determine an appropriate response. The assignment of job responsibilities based upon sex-role stereotyping is one example of a gender-based decision. Human services workers are just as capable of generalizing about a person's abilities as anyone. Workers may label a male who speaks up in a staff meeting "assertive" while his female counterpart is considered "aggressive." A male therapist may be assigned a male client because of the vague perception that the client needs a male. A fieldwork student may not expect a male supervisor to be outwardly caring or supportive, but that same student might expect a female supervisor to be continuously nurturing. Students and experienced staff alike may be tempted to act solely on the basis of gender-based perceptions. We all may find ourselves guiding a client or co-worker to be more or less "male" or "female" depending on how we have learned to view human beings. We must also remember that our clients will do the same with us.

Try to respond to the following questions as candidly as possible. The questions are not intended to produce specific correct or incorrect answers but rather to assist in self-exploration and contribute to discussion.

1. Define and describe those aspects of yourself that constitute your primary male or female qualities.
2. Identify positive qualities you possess that typically might be considered traits of the opposite sex.
3. Identify the primary sources of learning for your own gender-based behavior.
4. Describe those features of "maleness" or "femaleness" that you believe are learned rather than biologically inherited.
5. Have any of your views about women and men changed over the course of time? Please discuss the factors that influenced your views.
6. Do you prefer working with men or women? If so, why?
7. Do you prefer having a male or a female supervisor? If so, why?
8. Have you experienced any greater or lesser difficulty working with male or female clients or co-workers? If so, what may have contributed to the experience?

Age-Related Perspectives

The ages of workers and clients will always be one strong determinant of their relative perspectives and of their experience in the helping relationship. The experiences that we all have accrued over time influence a broad spectrum of behaviors and attitudes. We generally expect that differences will exist between members of distinct age groups. In fact, the larger the generation gap, the greater our expectation that differing attitudes will exist. This is not inevitable, but not altogether unlikely. Although we may value the wisdom that life experience brings, our culture regards younger members of our society with considerable admiration, for a number of reasons. Without any doubt, younger and older members of our society are aware of the distinctions between them relevant to age. How these distinctions are interpreted is a function of the individual and his or her system of attitudes and beliefs. Our attitudes concerning our age and age-associated qualities are directly related to how we behave.

Age, like socioeconomic status, religious affiliation, or gender, is a significant variable in the client-worker mix. Clients who are older than their assigned interns may question the ability of the interns to truly understand their concerns. Students are prone to similar feelings of anxiety when confronted with an older client. Over time, however, most students begin to realize that differences in age are no more salient than other areas of divergence between themselves and their clients. In fact, if we were to limit ourselves to working with populations that were identical in age or any other manner, we quickly would tire of the monotony. Differences stimulate learning and contribute to interesting interactions among parties; they don't necessarily impede communication. Interns often are surprised to learn that they were much more preoccupied with the difference in age than were their clients. Once an intern demonstrates a level of interest in approaching the task at hand, concerns about age seem to diminish. Of course, a number of students may be *older* than their clients, colleagues, or supervisors. In these instances the dynamics may be similar but reversed. As mentioned earlier, a student intern who is markedly older than those in positions of authority may have a more difficult time adjusting. Admitting feelings of apprehension or uncertainty allows the student to move beyond these concerns.

Some familiar dynamics occur in relationships between individuals who differ substantially in age. We often will be inclined to respond to those who are older than ourselves as we have learned to respond to authority figures such as our parents. It doesn't take a trained clinician to realize that unresolved areas of conflict between ourselves and our parents tend to be replayed later in life with others who remind us of those earlier relationships. Remember, however, that these powerful thoughts and feelings are often just beyond our awareness. This is one reason that supervisors and other consultants are available to assist students in recognizing the dynamics that exist in all intern-client or co-worker relationships.

The following questions may help you gain insight regarding your own views on age and age-related issues:

1. Describe the client population of your setting with respect to age. How does your age compare? What problems, if any, does this create for you or for the clients?
2. What age groups do you feel most comfortable working with? Why?
3. What age groups do you feel least comfortable working with? Why?
4. What advantages or disadvantages exist for you relative to your age in your particular internship setting?
5. What do you like the most/least about your current age?

Gay and Lesbian Perspectives

Few topics of discussion elicit such intense reactions as those that concern gay and lesbian individuals. As a society we are not presently in a state of uniform acceptance or even tolerance of homosexuals. There are many reasons for this. Religious and familial training have a profound impact on the development of attitudes concerning homosexual behavior. Views that have formed during a lifetime of interaction and learning often are deeply rooted and resistant to change. We need only to consider the passionate debate regarding the presence of gays and lesbians in the military to realize the scope of dissention concerning this topic.

Although the purpose of this text does not include altering the fundamental attitudes of readers concerning homosexuality, it is important to point out that human services workers must expect to encounter all kinds of clients, including lesbians and gay men. Workers need to develop their capacity for tolerance and understanding without prejudice at all times, for all clients. No matter what one believes about the fundamental nature of sexual preference, the gay or lesbian individual probably does not seek the services of a human services professional as a means toward changing his or her core identity. In fact, the sexual preference of any individual by no means provides a complete description of that person. Many times clients withhold information about themselves from workers until they have the opportunity to ascertain whether they can fully trust the worker. There may be no area of greater sensitivity to individuals than that which concerns their relationships and sexual behavior. Gay and lesbian clients understandably may be cautious about sharing information about themselves, even with a human services professional, given the overall climate of our society.

Whenever we react vehemently to a particular aspect of an individual, it probably reveals important information about ourselves. Sexual orientation and sexual preference are but two of the many factors that influence how we perceive ourselves. Human services professionals have a responsibility to step back and examine their response whenever they have a strong reaction to a client's behavior. In this regard, the demands on human services workers differ somewhat from those in other professions. Our own views will overlap with our perceptions of our clients and play an important role in our professional interactions.

Answer the following questions, and consider what your responses reveal about your attitudes toward gays and lesbians.

1. Describe any reluctance that you anticipate a gay or lesbian client having about approaching your fieldwork setting for help. Indicate any ideas that you may have about improving the delivery of services at your location.
2. What was the general climate in your family of origin concerning gays and lesbians? To what degree do you now share similar views?
3. Do your religious beliefs influence your views concerning gays and lesbians? If so, how do you anticipate these views affecting your interactions with clients?
4. Are gay or lesbian clients better served by gay or lesbian workers? Please elaborate.
5. Rank the intensity of your feelings about this topic from 1 to 10, 1 being *mild* and 10 *intense*. How did you arrive at your decision?

Geographic Perspectives

Anyone who has had the opportunity to travel within the United States has observed the regional differences that are evident throughout our country. People are shaped and colored by their surroundings and their environment, including that which we refer to as their geographic area of origin. Admixtures of family, neighborhood, community, and extended regional influences combine to influence our unique development. Although we may one day relocate to areas far away from the communities of our childhoods, an indelible imprint probably has been made on us by the customs and beliefs that we witnessed and shared as children and young adults. This is not at all detrimental unless one determines later as an adult that particular influences are undesirable and should be discarded. We often may consider the regional aspects of our upbringing to be a most desirable, admirable mark of our uniqueness. Indeed the multiplicity of regional influences contributes to our heterogeneity and richness as a nation.

Whether one has a background of experience in a rural or urban area probably does have some influence on a host of thoughts and behaviors. There are a number of common comparisons that you may wish to consider when weighing the relevance of your individual scenarios. Were you raised in the deep South or in northern industrialized areas of the country? Did you grow up on the Eastern seaboard or in the Pacific Northwest? Perhaps you are a midwesterner or are of southwestern background. It may be that you or your parents emigrated from another part of the world long ago, or even quite recently. Obviously there are numerous possibilities, each with its own potential for influencing your belief systems and personality. Customs and communication styles do vary in different parts of our country; dialect, expression, and inflection of speech vary according to the region. Whether you are put off, amused, or delighted by these variations may depend on your own individualized set of expectations for social interaction. Take a moment to imagine the effects of your exposure to the environments that have been an integral part of your life to the present. How might you have been different if you had been reared in a totally different locale? If your parents were reared in different parts of the nation (or the world), what do they convey in their manners or beliefs that is an outcome of their unique backgrounds? Are you pleased or disquieted by these particular influences?

The following questions may help you learn more about your own geographic perspectives.

1. Describe the influences of your geographic area of origin upon your own unique communication style. Would you change anything if you could? Why?
2. In your experience, do particular dialects or regionally influenced styles of interaction pose a difficulty for you in communication? If so, why? What do you do to adapt?
3. Is there anything striking about your own style of communication that may pose difficulty to another person in communicating with you? Be descriptive.
4. How might your own clients describe your communication style? Would they be likely to infer a particular regional influence in your communication and interactive style?
5. Do you assume that a person possesses certain characteristics based on your identification of their regional influences? Is that a helpful or harmful process on your part?

Political and Philosophical Perspectives

A well-known adage says that individuals should refrain from discussing either politics or religion at the dinner table. It is not difficult to understand why such an attitude exists. Politics, like religion, can be a volatile subject in any social gathering. Whether you intend it or not, clients or co-workers may elect either to inquire about your political viewpoints or to share their own. In any event, you'll need to consider what it is that you feel comfortable sharing, because others are likely to form some judgment of you based on those disclosures.

Regardless of the degree of passion that you have for political issues, some topic in the political arena is likely to concern you. If this topic is raised in conversation, you would be less than human not to have some reaction. Like the other subjects discussed in this chapter, politics may generate emotional responses of considerable intensity. Most of us can recall at least one instance of co-workers arguing their convictions while gesturing emphatically and elevating their normally subdued voices. (You may have even been a member of the group!) Some topic or issue is likely to interest even those who might not typically initiate such a dialogue.

If a client or even your supervisor asks your opinion regarding a current political issue, take a moment to consider before you answer. Understand that whatever you say may lead to further commentary or exploration of your views. You may be asked to elaborate or even to defend yourself. Politics is a highly charged topic of conversation; because it overlaps with other topics such as law and health care, and also such emotionally laden topics as welfare and abortion, all roads may lead to places that you had no intention to visit. Admittedly, these issues represent some of our core societal concerns. Many of them belong in the province of social science and, by close association, the human services. However, they are not necessarily the focus of our everyday activity.

Are we suggesting that workers should avoid all mention of politics or related topics in their work with colleagues or clients? Of course not. No subject is by definition off-limits. Yet entry into these volatile areas should be made cautiously, with an appreciation of the possible implications. For example, if a client solicits the worker's opinion on a sensitive political issue and discovers that their views differ substantially, will he or she feel rejected or attacked? If the worker shares a political perspective that contradicts that of the client, will the client perceive the worker as insensitive or lacking empathy?

How should workers decide whether any expression of their views is appropriate in a given situation? Again, consider the implications and proceed cautiously. If you realize that a conversation is becoming especially heated and threatens to injure an otherwise stable relationship, you'll need to back off and take steps to diffuse the situation.

Now ask yourself the following questions:

1. What are the political opinions that you feel most strongly about? How do you respond to disagreement over your most personal beliefs?
2. Do you feel obliged to express your dissention when someone else describes views that are contrary to yours?

3. What individuals are responsible for shaping your political views? Do you feel responsible for defending those persons when they are criticized?
4. How flexible or open are you to differing political ideas? Are you more or less defensive in this arena compared to others? Why?

Perspectives Pertaining to Physical Challenges

For those individuals who have no serious physical challenge, the notion of what it is like to cope with and overcome such a disablement is difficult to comprehend. We generally take for granted our capacity to climb a set of stairs or move about as we choose. Anyone who has worked in a setting where patients are struggling with the loss of a physical function has witnessed the significant adjustment that this requires. It may be that we are not able to fully appreciate some of our gifts until we face their loss.

Many readers may be thinking at this point that their perspective changes very little when they encounter an individual with a physical challenge. Ask yourself this question, then: How many times have you seen someone who is missing a limb or is restricted to a wheelchair and felt yourself fortunate not to be that individual? We tend to have some degree of sympathy for or, alternately, feeling of advantage over, individuals who are challenged. Although this response may not be typical, some persons may be frightened or repelled by the presence of someone who is markedly different from them physically. Perhaps the presence of a physically challenged individual awakens fears of physical harm or loss—fears that are present to some degree in all of us.

Consider the perspective of the person who is in the position of being physically challenged. Quite likely, he or she does not wish to be viewed only in terms of their physical makeup. How would you like to be seen primarily as the sum of your physical characteristics? Most of us would be averse to that constricted perspective, preferring instead to be appreciated for the sum of all our human traits.

The manner in which we view our physically challenged clients certainly will be influenced by any exposure we have had to people with disabilities. In addition, our own physical well-being may affect how we view others in this regard. It is also important to understand that the presence of a physical challenge alone does not relegate an individual to a specific status or group. All individuals differ in infinite ways in spite of any one shared characteristic. We risk overgeneralizing when we examine any population in terms of limited variables. When interacting with a client who is physically challenged, we must remember to consider both our own impressions and the point of view of the client regarding his or her particular challenge. What we believe to be a minor limitation may in fact pose an extraordinary challenge for that person; conversely, what we perceive as a traumatic loss may contribute only slightly to the client's overall self-image.

Answer the following questions, which may help you clarify your attitude toward physically challenged persons:

1. Define what constitutes a physical challenge.
2. Indicate how your own physical challenges have affected your life.

3. How would you personally adjust to a major loss such as vision, speech, or the ability to walk?
4. Have you ever felt any discomfort in the presence of a physically challenged person? If yes, why do you think that occurred?

Educational Perspectives

One measure of value or prestige that often is assigned to individuals in our society is their level of formal education. It is no secret that, occupationally, highly educated persons tend to hold elevated positions compared to their peers. (Of course, education is only one variable that relates to occupational placement.) Educational level also is related to overall socioeconomic well-being. Most of you who are reading this text all share a common investment in the process of education. A relevant question, therefore, is: How do we tend to view people differently depending on their level of education?

Most professionals have an acute awareness of how their qualifications, including level of training, compare to those of their peers. It is likely that most interns feel somewhat sensitive about their placement at or near the bottom of the organizational hierarchy. How do you feel about being an intern and having less education or training than the other workers at your location? Do you feel at ease, or do you feel threatened, perhaps defensive? Any or all of these reactions may be normal on occasion. In any organizational structure, some ranking of power and leadership exists. Often, there is a power differential between client and worker as well, and a subtle form of discrimination may occur between worker and client related to educational level. This may occur in either direction. For example, a client may inquire as to the level of degree or licensure the worker possesses. Although this is certainly appropriate on one level, the client also may be trying to discern how similar the client and worker are in this regard. In turn, the worker may have some reaction to a client who has a markedly different level of schooling. Do we rank ourselves vis-à-vis our clients? We may answer a resounding "no!" because we hate the thought of imagining ourselves as better or worse than the individuals whom we would hope to treat as perfectly equal. Yet suppose you have one client who is a physician and one who is a laborer, a high school dropout. Do you honestly have different reactions to them? Even if your answer is yes, remember that having a particular initial reaction does not obligate you to act in a predefined manner. Just be aware of those thoughts and take time to reflect on them.

Differences in education and training among workers and clients color the relationship in a variety of ways. Clients who have relatively greater education and also well-developed verbal skills may be comfortable working with a college-educated helper in a context where talking is stressed. In turn, workers whose clients have an educational background similar to their own will likely perceive those clients as easier to help. (It's possible, though, for a worker to experience some anxiety in the presence of an articulate client whose verbal-relational skills surpass that of the worker.) Have you ever heard a therapist suggest that a particular client is "well suited" to counseling? Perhaps what he or she means is that the client is perceived as able to relate well to the therapist.

Do not forget that nonverbal communication can be as important as spoken language. Workers or clients who feel uneasy are likely to show it, even if they do not articulate it. Remember, too, that a worker's degree of comfort is likely to be related to his or her overall self-confidence and self-esteem. A worker whose confidence and self-esteem are high will be more effective in dealing with any client.

Take a few minutes to consider the following questions:

1. What effect does the educational level of a client have on your own feelings of comfort in the relationship?
2. How satisfied are you with your present level of formal education?
3. When questioned by someone about your level of education, how do you react?
4. What were you taught in your family about the value of education or about how a person's worth relates to his or her education? Does that early training influence your views?

Ethnic and Familial Perspectives

Throughout this text we have emphasized how one's life experience will significantly modify and affect both behavior and perception. Without question, our family upbringing has momentous impact on our interactions as an adult. Although we all are human beings, we possess profound dissimilarities with respect to our families. Because one function of families is to socialize its members, people from different types of families will have divergent patterns of interaction. Let's consider a few of the dimensions that may be examined in an attempt to describe the structure and functions of families. Families may be analyzed in terms of how they are organized: by structure, ethnicity, composition, and behavior. Families have many complex tasks to perform, and they vary in their effectiveness in carrying out those tasks. Persons may recall a wealth of pleasant memories from their family of origin, or they may experience considerable pain as a consequence of those recollections. Most often, there is a distinctive mixture of experiences that combine to form the basis of our original family life.

The social sciences collectively help to define the cultural/ethnic/familial distinctions that exist worldwide. Your effectiveness and sophistication as a worker will be enhanced to the degree that you are willing to learn about your own history as well as other backgrounds and cultures. Consider the following list of family functions, features, tasks, and other descriptive qualities that may apply to a given family. Because this list is not exhaustive, we encourage you to think of other salient features of family life that contribute to a more complete listing. As always, a dialogue among students will help to bring this material to life while enriching and expanding your understanding of this important topic.

Descriptive or Structural Features of Families

Level of closeness or disengagement of members
Level of organization or chaos
Socioeconomic status
Clarity of boundaries between members
Composition/makeup: single parent, intact, blended, and so forth

Predominant language(s) taught
Inclusion or exclusion of extended members: grandparents, aunts, uncles, and so on
Religious beliefs and practices
Dietary habits and customs
Level of support provided to members
Level of safety provided to members
Cultural and ethnic practices, including influences from differing locales, regional or worldwide
Capacity to appropriately socialize and control members
Ability to assist members in a process of individuation and development
Ability to develop progressively as a family organism
Flexibility or rigidity
Ability to respond to stress
Morals, ethics, and values taught
Amount of role-specific or gender-specific instruction
Use of nonverbal communication
Instruction relative to affective (emotional) expression
Ability to ensure the survival and well-being of its members

All of the foregoing comprise our collective notions of what we come to consider normal in the context of family functioning. What was the quality of your family life? It is through that lens that you will view others with whom you interact privately and professionally.

The following questions may help you explore your own views on family functioning:

1. Which of the above dimensions would you consider to be the most crucial aspects of family functioning? How did your family of origin operate in those areas?
2. Are you sensitive to particular aspects of family functioning? How is that a possible outgrowth of your own family history?
3. If contact with families is a part of your internship, describe your experience to date. For example, has it been challenging, anxiety provoking, or . . . ? Why?
4. What do you plan to incorporate in your family life that was not a feature of your family of origin? Please elaborate.
5. What do you plan to eliminate from your family life that was a prominent feature in your family of origin? Please elaborate.
6. Is there a family structure or type that is particularly unfamiliar to you or hard to relate to based upon your experiences to date? How do you plan to manage your reactions?

VIGNETTE ONE

A municipal shelter that houses minor children and adolescents who are in temporary placement has several children in residence who are hearing-impaired. These residents are not separated from or treated differently than the general population of children, which is very diverse. The treatment staff consists of a normal complement of social workers, psychologists, and one human services intern. The intern enjoys a wide range of duties, including daylong interaction with the child residents and supervision of their normal activities and mealtimes. Overall the setting is considered to be challenging as well as rewarding; considerable demands can be made on a single worker in a moment's notice.

On one particular afternoon, the children were assembled for a routine trip to the cafeteria for a regularly scheduled lunchtime. Along the way, the staff became aware of some nonspecific rumbling among the residents but thought little of it, as some degree of disruption and limit-testing was not out of the ordinary. Upon arrival in the cafeteria area the level of noise and activity in the group seemed to escalate, especially at one particular table that included a number of the hearing-impaired residents. The intern responsible for that group of children attempted to act quickly to restore order and avoid any further intensification of behavior. It soon became apparent that the student was having difficulty managing the situation; the residents seemed to ignore all of her attempts to intervene. One after another child was grabbing food, shoving peers aside, and clamoring for different items on the table. When the student raised her voice or signaled with her hands for attention, the residents persisted in ignoring her efforts and intensifying their own actions. According to a nearby staff member, the children seemed determined, frustrated, and angry. They seemed to be out of control and were acting at times like they had very little training in mealtime social behavior. Eventually, the intern and other staff members used physical restraint to separate and calm the distraught children. In the end, many of the children had to forgo lunch, and the intern was left feeling confused and irritated by the reactions of the children. It was unclear why the children had become upset to begin with and why the intern's attempts to intervene had been so ineffective, or perhaps even inflammatory.

QUESTIONS FOR DISCUSSION

1. What limitations were inherent in the setting that may have contributed to the acting-out of the children? How would you address them?
2. Identify the alternate courses of action that may have been considered by the intern in this vignette.
3. Describe what the intern might do subsequent to this sequence of events to help remedy the outcome of the interactions. What further training is warranted?
4. Indicate what emotions and thoughts the children may have been experiencing. Why did they behave as they did?

ANALYSIS

Sometimes it is not easy to interpret behavior, especially in a stressful or highly charged context. Behavior that might ordinarily be understood may be misperceived when one or more variables differ from those that are commonly expected. In this situation, miscommunication occurred from the outset as the student and others present struggled to comprehend the meaning of the children's behavior. Was the original labeling of the behavior as acting-out correct? The student's attempts to set limits were seemingly based upon a premise of misbehavior that served only superficially to describe the dynamics present in the group. Consider what may have been missing in this scenario. What special or descriptive features of this population existed that might help us interpret these events? Certainly the existence of deafness in these children may have been overlooked. And anyone who has had the pleasure of working with children knows that acting-out often represents more than simple misbehavior. If all behavior is communication, what was being communicated by the children in this setting? Consider several possible interpretations of the effect of deafness on communication. What is the relevance of this special challenge to a child in this kind of setting? It may be helpful to imagine yourself in a similar circumstance or, better still, to recall any related experience that you have encountered in the course of your lifetime. What were the emotions that you felt?

VIGNETTE TWO

A nursing home community providing residential care for the elderly is the setting for a wide variety of activity, including regular visitation by family members. Students assigned to this facility provide direct services to residents throughout the week in order to have exposure to most scheduled activities. As a result, students witness a number of family interactions that include the emotional interplay that occurs between members on an ongoing basis.

During the course of family visitation, relatives and friends often express the concerns that they have been unable to voice during times of separation from their loved ones. On one particular afternoon, visitation was in progress between families and friends and interns were free to move about the facility to assist and interact as needed. In general, the events were typical and quiet, as visitors moved throughout the various areas. One intern on duty was a student in her mid-thirties, of Asian-American background. As she entered a large common area containing several families, she became aware of one particular elderly resident and a middle-aged woman who appeared to be her daughter. The two were engaged in an angry exchange, although the daughter seemed to be the more verbal of the two. After watching for several minutes, the intern noticed that the daughter was in fact quite brusk in her actions toward the mother. On a number of occasions the younger woman cut off the parent in conversation and gestured pointedly with her hands. Eventually the daughter ended the meeting in a rather abrupt fashion; she raised her voice in a final outburst that seemed contemptuous of the elderly parent, who was more resigned than resistant. The intern felt helpless and was upset by the actions of the daughter, whose behavior she considered inappropriate and inconsiderate. The student left her fieldwork later that day determined to speak to her supervisor about the wisdom of allowing the daughter to have contact with the elderly resident.

QUESTIONS FOR DISCUSSION

1. Discuss the possible contribution of the intern's ethnic background to her interpretation of the interactions she witnessed.

2. Describe a hypothetical family background for the student that might explain her feelings about the mother-daughter relationship.
3. Indicate your ideas about possible facility needs for multicultural education and training for staff and students. What specific areas would you emphasize?

ANALYSIS

Any outside observer might interpret the interactions described in this vignette in a manner similar to that of the student. To someone who is not aware of preceding events or the emotional state of the participants, the origin of an apparent conflict is unknown. In addition, the viewer brings his or her own perspective to bear in any interpretation. Here, we are aware of the outward irritability of the mother and daughter, but (similar to the intern) we do not have the advantage of knowing anything more about them. If the student is correct in her assessment of the situation, what can be further understood about the precipitating events or prior history?

In this case we must surmise that some level of frustration in one or both of the family members was generated by their interaction or perhaps was present prior to their meeting. The mother may have any number of reasons for being upset, including concerns over health, family problems, financial worries, or sadness over the loss of a close friend. Alternately, the daughter may have her own reasons for being angry, which may include anxiety about her mother's health or perhaps feelings of guilt or helplessness relating to her parent's admission to the facility. Sometimes anger is a sign of deeper, more difficult emotions.

The student, to her credit, appears to be genuinely concerned about the welfare of the elderly mother. We also can discern some strong feelings about the manner in which the daughter and mother related, which likely are influenced by the intern's own life history. Together, the student and supervisor need to examine not only the interplay of the mother and daughter, but also how the student's perspective contributed to the interpretation of the events.

SUGGESTED READINGS

Buhrke, R. A. (1989). Incorporating lesbian and gay issues into counselor training: A resource guide. *Journal of Counseling and Development, 68*(1), 77–80.

McGowen, K. R., & Hart, L. E. (1990). Still different after all these years: Gender differences in professional identity formation. *Professional Psychology: Research and Practice, 21*(2), 118–123.

McGrath, P., & Axelson, J. A. (1993). *Accessing awareness & developing knowledge: Foundations for skill in a multicultural society*. Pacific Grove, CA: Brooks/Cole.

Morrow, K., & Deidam, C. (1992). Bias in the counseling profession: How to recognize and avoid it. *Journal of Counseling and Development, 70*(5), 571–577.

Pederson, P. (1988). *A handbook for developing multicultural awareness.* Alexandria, VA: American Counseling Association.

Ridlen, S., & Dane, E. (1992). Individual and social implications of human differences. *Journal of Multicultural Social Work, 2*(2), 25–40.

Sue, D., & Sue, D. (1990). *Counseling the culturally different*. New York: Wiley.

Vacc, N. A., Wittmer, J., & DeVaney, S. B. (1988). *Experiencing and counseling multicultural populations* (2nd ed.). Muncie, IN: Accelerated Development.

Chapter Six

Surviving in the Work Setting

- Learning to Solve Problems with Flexibility, Creativity, and Tolerance
- Maintaining a Connection to Others
- Understanding Conflicts in the Workplace
- Self-Care Issues
- Vignettes
- Suggested Readings

Learning to Solve Problems with Flexibility, Creativity, and Tolerance

Human services professionals encounter a vast range of problems on a daily basis. Some situations are mundane and straightforward; others involve complex scenarios that demand considerable concentration and expertise. As students, you are not expected to manage the intricacies of your assignments in the same manner as a seasoned veteran, yet you very well may face the same types of dilemmas that confront the most accomplished professional. Not surprisingly, experienced staff tend to exhibit greater composure and confidence. Their self-assurance is the outcome of training that leads to a sophisticated way of approaching various tasks.

The Professional as Problem Solver

By definition, human services professionals are specialized problem solvers. Clients and others who interact with them expect to benefit from their level of expertise. This expectation adds to the pressure to succeed that many workers feel.

One problem trainees must deal with is the absence of sufficient time to master those skills required to address the full range of concerns brought to their attention. Many professionals can recall confronting pressure-packed situations that had not yet been covered in classroom discussion. Nevertheless, they still were expected to offer significant and meaningful help. Mastery of the content of one's academic discipline is an important goal. It is, however, only one aspect of professional development. Experienced staff who successfully manage stressful assignments did not develop their skills overnight or by accident. Training, exposure to senior staff, effective supervision, and even some measure of trial and error all contribute to a professional and proficient approach. Time and training allow didactic and experiential learning to mesh, allowing a worker gradually to establish a sound professional identity. Although no one style conclusively defines the ideal professional, we can identify several features common to experienced and effective workers.

Flexibility and Creativity: Key Components

Developing professional skill involves a process of growth in many dimensions. Fortunately, success breeds continued success as one's confidence builds. Time and the opportunity to take on progressively more difficult assignments also helps. Conversely, failure seems to diminish personal confidence. Human services work is by no means an exact science. Individuals do not act or react as we may sometimes predict, no matter how expert we become in our work with them. It is important to maintain an outlook that allows you to grow and learn without demanding perfection. Realistically, interns can expect to make mistakes. When experienced professionals make mistakes, however, they are far less critical of themselves than are most interns. When you make an error in judgment or action, it is important to turn to your supervisor for support and guidance. He or she can help you learn not to repeat the same mistakes over time, in similar circumstances.

Effective professionals are flexible rather than rigid in their approach to human predicaments. Clients usually present concerns that have been the focus of repeated problem-solving attempts over a period of time. Often, one striking feature of these attempts is the repetitive and narrow range of ideas they represent. One task of human services workers is to assist in identifying a greater range of solutions as well as to enhance the flexibility of the system (micro or macro) to allow for improved problem solving in the future. Professionals have some advantage due to their perspective from outside the client's system. This perspective assists the worker in helping clients to consider the range of available options. Workers should also examine their own willingness to be flexible rather than rigid, creative rather than unimaginative. Human services professionals do not learn a body of concrete rules or laws that can be applied uniformly to all human problems. A willingness to adjust, adapt, and devise creative solutions is a critical component for success for clients and professionals alike. The successful manager, teacher, therapist, and student intern all share a common need to be as inventive and resourceful as possible. One practice that often gives birth to innovative ideas is brainstorming, a technique that involves the spontaneous contribution of ideas from all participants. Students also may want to examine the styles of creative and productive individuals from various fields, and not limit themselves to ideas that arise solely from their own discipline.

The Willingness to Negotiate

Consistent with the ability to maintain a flexible and creative posture is the willingness to negotiate with others. Human services workers often are called upon to enter into negotiation. The individuals and groups that workers interact with typically are experiencing some form of conflict, so managers, teachers, and therapists learn to engage in conflict resolution as a consistent aspect of their work. They often must broker solutions between parties who are at odds with each other and who also may express disagreement with the workers themselves.

Whether or not you relish conflict or disagreement, you must anticipate that pursuing a career in the human services means you will experience discord firsthand. Ask yourself what thoughts and feelings you typically experience when you are in the presence of people who are disagreeing or perhaps vehemently arguing strongly held beliefs. What do you commonly think or feel when someone directly challenges a personal perception that you believe to be absolutely correct? Remember, you serve as an example to others, who view you as a model that they should strive to emulate. Rigid, dogmatic insistence on being right or on consistently "winning" in exchanges with co-workers or clients is not likely to earn you the respect of colleagues or clients. The willingness to negotiate is an attitude that precedes successful dialogue.

The Importance of Tolerance

Human services professionals continually are exposed to problems, cultures, and settings that are unfamiliar. Inevitably, individuals and groups that the intern encounters will challenge that intern's tolerance of individual differences. Even the most sensitive or astute individual may react to some individuals with fear or disapproval. Similarly, a client or co-worker may be uncomfortable with personal traits of the intern. It is naive to expect to be compatible with all the individuals that you'll have contact with over the course of your internship or career. You need to ask yourself whether the differences you perceive are a function of religious, educational, familial, or cultural background. Do prejudicial beliefs play a role? Once you become aware of the reasons for your responses, you can begin to take steps to expand your level of tolerance.

The fieldwork setting also can give rise to feelings of disgruntlement regarding the physical surroundings. Although some interns may be fortunate enough to enter a setting that is comfortable and attractive, others may inhabit quarters that make them feel cramped, ill at ease, or simply overwhelmed. A large county hospital may offer a bewildering array of corridors, offices, and labs, and weeks may pass before an intern is fully oriented. In addition, the physical location of the internship setting may or may not replicate the neighborhood or community most familiar to the student.

Interns are confronted with the need to examine and expand their level of tolerance in many different areas. Taking the view that this is an opportunity to increase your capacity for adapting to challenging and unsettling circumstances will allow you to develop greater tolerance toward all aspects of your job. Actively pursue opportunities to broaden your knowledge of people, ideas, and settings outside your personal frame of reference. Education combats fear and ignorance, and increased tolerance often emerges.

Maintaining a Connection to Others

Avoid Isolation

As we have discussed earlier, one means of limiting the amount of stress you experience in the workplace is to remain in contact with other individuals who can provide support and reassurance. Some individuals may tend to isolate themselves when they are under increased pressure. If you are prone to this response, force yourself to interact with other staff and students on a regular basis so that you will feel a part of the organization.

It may be easier to become lost in large, unfamiliar settings than it is in smaller, more discrete locations that encourage frequent and close collaboration. For some students, solitary study may be the norm. You may have to remind yourself to step out of your office in order to meet other people! It can be difficult to alter a pattern of isolation once it is firmly established, so make an effort to introduce yourself to others and begin to build relationships early in your internship. Students who approach other staff and participate in team discussions or even casual conversations may be the ones who are best remembered for a particular project or opportunity—simply because they made themselves known.

Remain Open to Input from Others

Although this may seem like a basic principle of successful interaction, it is amazing how often managers in organizations violate this notion. Staff members know it is typical, in large and small settings alike, for administrators to alter policy and procedure without seeking or accepting any input from those most affected by them. Similarly, individual members of human services settings sometimes forget that they are members of a larger social network. The advice of colleagues is nearly always available for the asking. Of course it is not always necessary to consult when making routine or simple decisions. To do so might convey a sense of helplessness or an inability to utilize one's own judgment. However, many of the problems faced in human services work are complex issues involving intricate technical, legal, or ethical dilemmas. Consultation never forces an individual to pursue a particular path, but it often reveals unrecognized possibilities. Students have the most consistent opportunity to integrate alternative perspectives about their work through supervision. Consider how open you really are to ideas and suggestions, particularly when they may not agree with your original ideas or plans. Do you tend to become irritated or defensive when a colleague or supervisor hints at an alternative strategy? or when you are directly confronted or challenged? Most of us have at least some occasional resistance to another person redirecting our activity, particularly when we are invested in the behavior for intellectual, emotional, or other personal reasons. To the extent that we reactively resist input, particularly as professionals in training, we close off opportunities for growth and professional development.

Enter into Dialogue with Colleagues and Co-Workers

Even the most experienced practitioner can become myopic when examining a difficult problem. Our training and life experience, although valuable, can cause us to view individuals and their problems in ways that restrict the development of creative ideas. In order to discover all possible solutions, we must enter into dialogue with others when facing complex issues. For example, health practitioners are expected as a matter of sound practice to solicit second opinions when faced with complicated treatment decisions. Similarly, other human services specialists should engage in interactive dialogues with other professionals as a matter of routine. This entails entering into purposeful exchanges with others with the intent of arriving at ideas that are more well developed and multifaceted than would otherwise be the case. Dialogue that is mutually respectful, lively, and open to divergent input is almost always beneficial. It can be a pleasure to collaborate with others to find solutions to difficult problems. And, failing to dialogue with others may mean that you remain unaware of potential obstacles or viable solutions. It's important that you learn how to approach another student, professional, or your supervisor, and that you become skilled in initiating and facilitating dialogue. The following questions may help clarify your ability to consult with others:

1. When was the last time you asked a colleague for his or her opinion about a problem at your work or internship? How did you respond to the input?
2. Do you think others view you as approachable or as a valuable source of ideas? If not, how could you improve their impressions?
3. Name several talents and skills that make you a helpful consultant.
4. How willing are you to dialogue with another person, instead of just offering advice?

Understanding Conflicts in the Workplace

Conflict: A Normal Process

If you have worked in any setting for a period of time, then you also have witnessed conflict among the participants. Conflict is expectable and can be anticipated in any group process. All of us have participated in some form of disagreement, dispute, or misunderstanding with our colleagues. Rather than asking whether conflict exists, it is more constructive to consider how much conflict is present and the mechanisms by which it can be resolved.

As we have indicated, the workplace is a social arena where interns are expected to interact with many different individuals. The diversity one encounters can be stimulating and also challenging. For example, you may be asked to share an office with a fellow student who is thoughtful and supportive—or with one who is a source of infinite frustration! (Close working quarters may bring out our best, or our worst, personality traits.) You may discern that most of the staff members are creative and dynamic individuals; however, some may be burned out and resentful. Because you cannot choose to interact only with those persons with whom you are entirely sympathetic, it's inevitable that some conflict will occur.

The Employee as Family Member

If you're getting the idea that the workplace is a little like a family, you're right. Most of us can recall instances of conflict in our own family life, no matter how healthy its composition. To some degree there is an unavoidable element of family group dynamics in most work groups. All of us carry forward the relationship styles we have learned over many years. Our behavior toward co-workers and administrators often is comprised of reflexive, habitual responses that have developed over time. Although our own interpersonal style may not be easy to assess, we can observe the ingrained tendencies in our colleagues' personal styles. For instance, one person may have problems relating to anyone in authority; another may appear incapable of saying no, no matter how extreme the request. Ask yourself how you relate to members of your own team, workgroup, committee, or department, and take note of predominant behaviors. For example, if you have a history of significant passivity in the face of conflict as an outcome of adopting a certain role in your family, you may anticipate a strong tendency to behave similarly years later, in the midst of a work dispute. Alternately, if you are comfortable with a dominant, directive posture, then you may tend to adopt a similar stance in your work. In either scenario, we advise you to analyze the strengths and weaknesses of your behavioral training with respect to your present-day context.

The Employee as Group Member

All groups have certain elements in common. We can describe them in terms of a number of factors, including style of organization, identifiable leadership, defined purpose, clarity of boundaries, morale or mood of the members, investment of power, homogeneity, and so on. And, groups have many complex functions. Any student or employee who becomes affiliated with a particular work site typically becomes a member of multiple groups. He or she is affected by the nature and quality of those groups and contributes to their unique makeup. Consider the variety of groups in which you participate during the course of a workday. For example, you may take part in a morning supervision meeting with other interns; later you may join a workgroup focusing on specific projects; and after that you may attend a staff meeting that is followed by lunch with several students and coworkers from other departments. In each of these particular groups you may view your role as either central or peripheral. You may be interested or bored, and you may initiate activity or follow the lead of others. Your ongoing experience will depend in part on how you participate over time. Think about the groups to which you presently belong. Which of these provide the most satisfaction? What distinguishes them from others that seem uncomfortable or unsatisfying? Perhaps your style is more suited to one type of atmosphere than another. It may be helpful to try to identify the interpersonal style that you exhibit in most settings. For example, are you shy and quiet? Do you tend to withhold comment in order to avoid embarrassment or conflict? Or are you perhaps so aggressive or dominating in your manner of speaking that other members feel put off by your presence? Human services work is generally very social in nature; workers are required continuously to engage in interaction with other individuals. A willingness to examine and modify how you function as a member of a group will enhance your effectiveness no matter what your area of specialty may be.

Here are some questions to consider as you analyze a particular work group and your own membership style:

1. What are the goals of this group?
2. Is there a clear structure with regard to leadership, procedure, schedules, membership, duties, purpose, and so on?

3. How cohesive is the group?
4. Is the group valued by its members or by the larger organization?
5. How clear and productive is communication within the group?
6. What special talents or skills can you offer the group?
7. Do other members value your participation? If not, why not?
8. Are you as active as you would like to be in the group?
9. Would you like to change anything about your role or style of interaction?
10. Who can serve as a resource to you when you have problems or concerns about the group?

The Importance of Assertive Communication

An assertive style of communication is learned. It develops over time, is the result of practice, and plays an integral role in one's self-confidence. As an intern, it may not be comfortable or easy at first to assume an assertive style when you practice among professionals who have the benefit of considerable training and experience. All the same, it is worth the effort to cultivate assertive communication skills, even when they don't come naturally. In fact, the earlier that you begin the process the better, since roles in the workplace can become harder to modify once they become established. It's hard to suddenly gather the courage to speak up or take an active role if you've been silent for several weeks.

As an intern, you'll be called upon to present ideas, participate in committees, problem solve, or actively intervene to resolve a crisis. Over and over again, effective verbal skills will prove instrumental in handling these situations effectively. For those of you who are somewhat reticent, the process of becoming an assertive speaker may be a bit harder, but it is by no means impossible. An assertive style is learned and acquired, not inherited. Observe experienced professionals in your setting who provide examples of strong, confident verbal expression. Be cautious in your comparison to other individuals. The interactive style or approach of someone else may not always be an appropriate model for your own behavior. Remember that a professional who has practiced for many years has probably acquired his or her speaking skills a little at a time. Because we all start from different points, goals should be individualized and realistic.

A core concept in developing an assertive communication style is to recognize when you feel uncomfortable. Feelings of discomfort or upset often are signals that something is not right. If you are irritated or angry, don't ignore it. Perhaps you are confused about an assignment or feel as though one more assignment or project will cause you to feel exasperated and overwhelmed. Perhaps you had an interaction with a client or co-worker that was frustrating or disappointing. All of these circumstances (and many others) call for you to speak up rather than remain silent. We can all think of circumstances where we know that acting assertively would have served us better than a passive approach.

Those around us often are in a position to view our behavior with some detachment and objectivity. If you believe you need help in developing an assertive style, your fieldwork supervisor can offer valuable support and input throughout your internship. Other staff also may serve as a source of assistance. In fact, other professionals may relish the opportunity to make a suggestion, offer feedback, or demonstrate alternatives.

A considerable amount of resource material is available for those who wish to develop their assertive techniques. Self-help literature can assist you in understanding a reluctance to be assertive, and offer you concrete advice about skill development. If progress comes to a standstill, we advise you to pursue counseling. Individual as well as group counseling may suggest additional strategies and provide a significant source of relief. Beyond that, the use of role-playing and videotape for viewing and rehearsing behavior can be helpful.

Ask yourself which of the following statements apply to your own ability to be assertive:

1. I never am reluctant to express my own opinion, even when others disagree.
2. I generally keep my feelings to myself.
3. The idea of disagreement with others makes me nervous.
4. I have a hard time saying no, even when I intend to.
5. I believe that it is my right to have my own opinions in all situations.
6. I'd have a difficult time as a leader in a group.
7. I'm not uncomfortable imposing limits on others.
8. I can refuse a request when I believe it to be too big or too unreasonable.
9. I speak up for myself when wrongly criticized.
10. Most people who know me consider me an assertive person.

Self-Care Issues

Keep Your Eye on the Big Picture

Internship responsibilities have the potential to become all-consuming. There is an unending stream of tasks to be completed, and before you know it you may be burning the midnight oil. Without intending to, you may find yourself working well after your peers have gone home or, worse yet, coming in on Saturdays on a regular basis. As resentment builds you may begin to wonder if you've made the right career choice. (It doesn't help that you receive no real pay in return for your long hours—a paycheck would at least help to offset some of your fatigue.) Being a beginner is not always a lot of fun. If you find that your world is becoming defined by the four walls of your office and the projects and meetings that are scribbled on your calendar, it may be time to back up and regain some perspective.

Considering the fact that you'll experience plenty of low points as well as periods of relative accomplishment, it's important to keep reflecting on the goals that you have set for yourself. The internship is but one facet of your training; it is not the defining experience of your career. Perhaps a subsequent internship will be even more satisfying than this one. You may decide to explore a different path within the field or take a different direction altogether. For some students, participation in the initial internship may solidify their commitment to a chosen area; others may be prompted to reconsider their plans.

Some students undoubtedly will be placed in settings that prove to be a poor fit. In spite of everyone's best efforts, the internship may not live up to the intern's expectations. Although you can and should make every effort to maximize the benefits of your fieldwork experience, some factors may be beyond your control.

Remember that the internship experience is only one aspect of your training. The skills you learn as a human services worker are accumulated over time and integrated into a whole person. No matter what the present holds, take a deep breath from time to time and enjoy the journey.

Ensure Your Self-Preservation

Like long-distance runners, interns must continually adjust and adapt to changing conditions while keeping an eye on the finish line. An overall goal of all interns is to successfully complete the entire placement, learning as much as possible along the way. In order to function optimally from beginning to end, it is necessary to attend to your well-being as a person at all times. It is easy to become discouraged or feel too tired to continue over the course of a long academic year. Multiple demands and projects may continue nonstop, regardless of your energy or enthusiasm. Don't take it personally. There really is no conspiracy to wear you down; this is just the nature of the pursuit you've undertaken. Later, when you have joined the workforce, similar pressures will exist day in and day out. Those who survive as professionals in the field of human services learn how to pace themselves. They draw on their personal resources and recognize their need to stop and rest from time to time. As an intern, you need to develop your ability to monitor the internal cues that tell you when you need to pull back, slow down, or take time for yourself. Because we all are individuals, no one specific rule applies to everyone. For example, ask yourself whether you usually can function effectively on only a few hours sleep or whether such a strategy should be reserved for emergencies. What about your eating habits? A regular routine of enjoying healthy meals may keep you from becoming exhausted, irritable, or run-down.

One significant challenge for students in placement is finding a way to structure and manage numerous responsibilities. It's important that you learn to attend to projects that are in need of immediate attention and defer others. An individual can accomplish only one activity at a time. Although this may seem obvious, it is surprisingly easy to become overwhelmed by trying to accomplish too many things in too little time. If you start feeling irritated or confused, that's a sign to slow down or stop momentarily in order to regain your focus and your equilibrium. Sometimes merely stopping for a break can help you to reevaluate and decide on a new course of action.

Although you of course will need to devote a considerable amount of time to your internship activities, don't neglect the importance of a life away from study and fieldwork. Some recreational activity and time spent with others is vital in order to maintain some perspective regarding the relative importance of work in your life. We all differ with regard to our specific needs for play, work, and social activity. Without some measure of each, however, we risk depleting our energy and motivation and becoming emotionally drained. Try to structure time for recreation and relaxation just as you plan and allocate time for study and practice.

Ponder the following questions:

1. How do I recognize the signs and symptoms of stress in myself?
2. When I do recognize that I am negatively affected by stress, how do I react?

3. Do those around me tend to identify my stress-related reactions or symptoms before I do? If so, what can I begin to do to improve my own self-awareness?
4. What can I do *now* to improve my ability to cope with and manage stress?

VIGNETTE ONE

Mary is a human services student serving her first internship in a county department of social services. Her placement is designed to involve her in learning about programs for the frail elderly, including outreach, education, socialization enhancement, and preventive health care. Mary is in her mid-twenties and has recently developed an interest in issues involving the elderly, although she never has worked with this population and is unfamiliar with her setting or with the county social service programs. She also is unclear about the kinds of jobs that exist in this field. Mary's grandmother died about six months ago after a long struggle with cancer. The experience was emotionally draining for Mary because she was close to her grandmother, who had raised her. Mary was delighted to find this internship after an extensive search and was happy that she finally had discovered a good fit for herself.

A few weeks into the placement Mary had begun to settle in, although she had only one opportunity to meet with her supervisor. Before she could schedule another meeting, the site manager informed her that her supervisor was being transferred to another location and that she would have to wait for a replacement. When she inquired about the length of time she might have to wait, the manager assured her that there was no need to worry because a new supervisor would be interviewed sometime in the next two or three weeks.

Several days later an assistant administrator approached Mary and asked how she was doing. Tearfully, she recounted her frustration at having so many unstructured hours on the job while she awaited the arrival of a new supervisor. To her surprise, the administrator invited Mary to help develop and coordinate an outreach program for homebound seniors in the community. Mary expressed some of her ideas at a program staff meeting. Several staff members were strongly supportive of her plans. One of the longtime members, however, was openly sarcastic and critical whenever Mary spoke. After the meeting this worker confronted Mary and complained to her bitterly about an event that occurred weeks earlier, when Mary fielded a phone call from one of the worker's clients. The worker told Mary to stay out of her cases and stop interfering with matters that were not her direct concern. Mary left the setting feeling confused about the interaction and irritated by the day's events.

Although Mary was hurt by the remarks of the staff member, she decided to approach her again in an effort to resolve the source of their conflict. Bolstered by the encouragement of her fellow students, Mary approached her colleague one day in the hallway. Once again the staff member rebuffed Mary and continued on with her sarcastic comments about Mary's "naive" community organization plans. Mary wondered if the co-worker simply didn't like her (she was several years older than Mary) or if some other unknown factor was at play. She decided simply to ignore the staff member in the future whenever possible. Later, during a coffee break in the lunchroom, she thought she heard the staff member making critical comments about her to a cluster of other workers sitting nearby. Mary, who now felt very angry, considered approaching the group but hesitated, wondering if another confrontation might only make matters worse.

QUESTIONS FOR DISCUSSION

1. Based on your experience, do you tend to engage in conflict with others who have particular characteristics (age, gender, physical features, personality traits)? If so, why?
2. In your opinion, what are the critical skills needed in order to coexist peacefully with others in a diverse work setting?
3. How should the supervisor play the role of mediator in interpersonal conflicts? What if the participants ask to be left alone?
4. What lessons did you learn about conflict resolution from your family of origin?
5. Evaluate your ability to manage your anger during a heated disagreement with someone. Are you able to express your thoughts and feelings as well as you would like to? If not, what can you do to improve your skills?
6. When was the last time you were asked to collaborate with someone? How flexible were you?
7. When was the last time you needed to be assertive with another individual? Would you change anything about your behavior in retrospect?
8. How would you characterize Mary's behavior toward her co-worker? What lesson(s) might Mary have learned that could benefit her in the future?
9. How would you describe the co-worker's behavior toward Mary?
10. Should Mary's co-worker be reprimanded? If yes, why? If not, why not?

ANALYSIS

In some situations it may seem as though there are as many opportunities for conflict or misunderstanding as there are people. Forceful personalities and varying viewpoints may provoke individuals to become easily agitated by one another. Often it is difficult to identify the original source of friction within a group. External pressures, such as increased workloads, changes in the physical environment, threats to job security, or administrative reorganization, all increase the likelihood of friction among participants.

The characters in this vignette are all struggling with a host of strong feelings, all of which seem valid given their individual perspectives. Often, conflict between members of a group or between one or more groups has a developmental history that may yield the information necessary for truly understanding the viewpoints of the players. Imagine yourself as a supervisor newly assigned to this setting. You suddenly are confronted with the need to negotiate a resolution. Consider how you might attempt to gather sufficient information for arriving at an informed opinion. Can you recall a scenario from your own employment history that is somewhat analogous to this one? How would you evaluate the performance of your supervisor or manager in that situation. What would you have done differently?

Remember that no matter what the facts are in a given situation, there is always an overlay of emotional reactions on the part of the participants. Some responses may be prompted by memories of other, similar experiences. This of course can make it difficult to unravel a dispute and identify the salient sources of anger and upset, and calls for a sophisticated point of view on the part of the group facilitator. What is your experience with conflicts involving two or more parties? Do you often feel compelled to respond in a particular manner? At various times in your career you undoubtedly will be called on by a co-worker to take a position or intervene in a conflict between two or more individuals. What will be your response?

VIGNETTE TWO

Mel is an intern in an inpatient chemical dependency program for adolescents. The setting is a substance abuse treatment center. Although he was nervous initially, Mel has become more accustomed to the daily rhythm of the setting. Mel is 30 years old and was raised in a rural area of the state. The setting is located in an urban center and draws clients from many regions. The age of the clients ranges from 13 to 18.

Mel requested that he be allowed to actively participate in the rehab counseling groups, and he has been allowed to sit in. Gradually, his contributions to the group have been increasing. He has been surprised by some of the disclosures that he has heard; many of the teens describe serious drug use and histories of major conflicts with family members. Mel has a history of some limited drug and alcohol experimentation in his adolescent years, and he engaged in his fair share of power struggles with his own parents. In his early twenties, Mel often was told that he had some problems getting along with authority figures, but he feels he's left those problems behind him.

One of the regular aspects of treatment scheduled for clients and their families is a weekly meeting that involves several families who gather in one large room. It is aptly called "Multiple Family Night." During the course of these meetings, several members typically speak about a wide range of issues and then listen to the perspectives of their own family as well as members of other families in attendance. The meetings vary in intensity, with the atmosphere occasionally becoming quite explosive as the content and emotions expressed grow more and more provocative. On one particular evening, Mel listens to an adolescent boy describe his drug use and his perceptions of its impact on friends and family members. When he is confronted by several group members about the accuracy of his descriptions, the youth grows sullen and then suddenly combative as he perceives the other group members to be attacking him. Although Mel has not been a very vocal member of the group up to this time, he finds himself boiling inside as he listens to what he believes to be an unwarranted attack on the adolescent. He speaks up on the teen's behalf, defending him even more adamantly than he perhaps intended. Surprised by the sudden input, one of the staff leaders suggests that Mel pause momentarily in order to allow other group members the opportunity to speak. Mel becomes irritated at the suggestion and brushes off another adult member's comments that he may be responding too personally to the scenario. After the meeting Mel seeks out other group members whom he thinks might support his position. The next day he feels more uncertain about his response and wishes he hadn't been quite as outspoken.

QUESTIONS FOR DISCUSSION

1. How relevant is Mel's personal background to his reactions?
2. What might be some possible consequences of Mel's comments?
3. How might Mel (or any intern) prepare for his or her role in this setting?
4. What events or scenarios in which you interacted with other people stimulated a strong emotional reaction on your behalf? How did you handle it?
5. What could Mel do when he finds himself at odds with the members of his internship setting (either staff or clients)? Suggest alternatives.
6. Identify which responses made by this intern were *assertive* and which were aggressive or angry. How might you have responded in Mel's place?

ANALYSIS

Our ability to understand our reaction to another individual relates to our capacity to integrate complex information, allow input, and recognize when our response is idiosyncratic. To a significant degree, all of our perceptions are colored by the unique nature of our own personal history. The training we receive as human services workers should help us to reflect on circumstances and analyze events in a manner that precludes a strictly personal reaction. We need to develop an expertise in understanding all human behavior, including our own. One advantage of participating in social interactions is the opportunity to receive feedback from our peers. We can note whether we become angry when the input we receive is contradictory to our own viewpoint. Perhaps our strong feelings may prompt us to action or cause us to reflect on or reconsider our viewpoint. What do you do when your views are challenged by another individual? Do you require a cooling-off period in order to respond constructively?

Workers, interns, and clients all interact on multiple levels as they influence one another's behavior. As human beings we expect to have some emotional responses to the people we interact with on a daily basis. In addition, we are influenced by our overall mood and physical well-being. Similarly, stressors that exist in our personal lives will increase or decline, combining to affect our performance and level of tolerance on the job. During periods of relatively low stress it is easier to respond to problems calmly and clearly. What are the stressors and other factors that may be influencing the behavior of the intern in this scenario? Think about your own setting as well as the internship or employment settings that you desire to join in the future. What can you anticipate about the emotional climates of these settings? What can you do to prepare for the stress or demands you will experience in these settings? What do you plan to do to offset the impact of conflict, stress, and fatigue as you proceed through your internships?

SUGGESTED READINGS

Cameron, J. (1992). *The artist's way: A spiritual path to higher creativity.* New York: Tarcher-Putnam.

Cherniss, C. (1995). *Beyond burnout.* New York: Routledge.

Covey, S. R. (1989). *The seven habits of highly effective people.* New York: Fireside.

Farber, B. A. (1983). *Stress and burnout in the human service professions.* New York: Pergamon Press.

Langer, E. (1989). *Mindfulness.* New York: Addison-Wesley.

Langs, R. (1993). *Unconscious communication in everyday life.* Northvale, NJ: Jason Aronson.

May, R. (1983). *The discovery of being: Writings in existential psychology.* New York: Norton.

Moskowitz, S. A., & Rupert, P. A. (1983). Conflict resolution within the supervisory relationship. *Professional Psychology: Research and Practice, 14,* 632–641.

Nelson-Jones, R. (1990). *Human relationships: A skills approach.* Pacific Grove, CA: Brooks/Cole.

Reamer, F. G. (1992). The impaired social worker. *Social Work, 37,* 165–170.

Rice, P. L. (1992). *Stress and health* (2nd ed.). Pacific Grove, CA: Brooks/Cole.

Chapter Seven

The Larger Issues: Maintaining Perspective

- Children's Issues
- Homelessness and Poverty
- Violence
- Chemical Dependency and Substance Abuse
- Vignette
- Death and Dying
- Vignettes
- Acquired Immune Deficiency Syndrome (AIDS)
- Suggested Readings

One of the prevailing catalysts for students pursuing professions in the human services is the notion of making a difference in their communities and influencing the larger social context. As you develop your interpersonal skills and cultivate areas of special interest, you may become acutely aware that certain societal issues affect you more deeply than others. Perhaps you or members of your family have experienced particular economic hardships that have made you more sensitive to the societal conditions of poverty or homelessness. You may have been raised in an environment characterized by prejudice based on ethnic, racial, or religious differences. Perhaps you were exposed to the effects of catastrophic illness, or to addiction, or were the caregiver to a dying relative. All of us will encounter one or more of these challenges sometime in our lives.

As we struggle to become effective and contributing human services professionals, we devote our attention to many aspects of individual behavior as well as to the forces that motivate the behavior of social groups. We try to understand various systems of values and complex philosophical ideas. We explore the role of emotions in human life and examine the myriad ways in which they are expressed. We hope that this preparation will serve us well when we confront the serious issues that are an unavoidable aspect of our work. We know that we will face scores of human problems with histories that span generations and we hope to induce some measure of change as a consequence of our efforts.

In this chapter, we will focus on a few of the larger issues that form a regular part of the landscape for human services workers. These issues provoke mixed feelings in many fieldwork students, and we encourage you to engage in thoughtful exploration and spirited debate as you consider them. Social problems of this magnitude can often be frightening and overwhelming, both to students and new professionals and their more seasoned counterparts. Although one or another of these issues may have sparked your initial interest in a career in the helping professions, confronting these issues directly and seeing how they manifest themselves in the lives of real people can lead to a change in perspective. Often the first change to occur is when idealistic expectations give way to realistic objectives. This change in perspective is a necessary one. However, even realistic goals may come to be viewed as unworkable or impractical when experiences in the field are less than gratifying or when problems encountered provoke intense or prolonged feelings of helplessness. In such an instance, fieldwork students may begin to convince themselves that they will never be able to make any meaningful difference. Their motivation decreases and they begin a downward spiral that often results in the condition known as professional burnout. The brief discussions that follow present ideas and activities that may increase your self-awareness and your ability to pursue your goals in ways that will lead to a sense of satisfaction and accomplishment.

Children's Issues

Although not all fieldwork students will opt to train in child-related environments, it is likely that issues concerning children will occupy a place of central importance for many human services workers. Those who do not provide services directly to a population of child clients may influence the lives of children through community programs, policymaking, proposal development, and grant writing. Others may work with spouses and parents in parenting classes or homeless shelters, or by providing assistance to victims of domestic violence. We all share responsibility for advocating responses to the needs and welfare of this most vulnerable segment of our society.

Confronting Child Abuse

The area of child abuse invokes tremendous discomfort in the hearts and minds of helping professionals. As a fieldwork student, you probably hope to protect children from any kind of abusive experience, yet you may also dread confronting this issue directly. Ambivalent feelings like these are endemic to this social problem, and even seasoned professionals struggle when faced with heavy responsibilities such as those associated with the legally mandated reporting of abuse.

Imagine yourself as a mandated reporter in a case involving physical evidence or a disclosure of child abuse. Use the following questions as a basis for discussion with other fieldwork students.

1. How familiar are you with the reporting statutes in your state?
2. How would you determine if a particular situation must be reported?
3. If a child has just disclosed information about abuse to you, what concerns might you have about your relationship with this child if you make a report?

4. How might you tell young children that you need to take steps to protect them?
5. What fears might you have in executing the duties of a mandated reporter?
6. What types of child abuse situations would be most difficult for you? Why?
7. How might you respond to parents who are angry with your decision to report their abusive actions?
8. Do you agree with the policies of your state concerning reporting child abuse?

It is comforting to imagine that when we are confronted with child abuse we will be able to think clearly, assess the problem correctly, manage our emotional responses, and follow through with our reporting responsibilities competently. Often, however, child abuse situations contain many ambiguous elements that make assessment and reporting difficult. And the serious nature of the issues involved can give rise to strong emotional reactions on the part of the workers. The first task you must face in an abuse case is the determination as to whether a type of abuse has occurred. Continuing education classes and coursework that focuses on the indicators, symptoms, and dynamics of child abuse give you an opportunity to enhance your ability to make a reasonable assessment. Developing skills in interviewing children about these difficult matters will help you to communicate with a child who otherwise may continue to suffer in silence. A large percentage of child abuse occurs within the family home, and a child rarely recognizes abuse when it happens. Children by and large do not question the actions of adults, particularly their parents, and even modern child-rearing approaches often reinforce this attitude in the name of "obedience." As a fieldwork student, you must be vigilant to physical bruises that are described as the result of an accident, to atypical sexual behavior or knowledge, and to other signs that may signal abuse.

To function at your best in matters that pertain to children, it is necessary to examine your attitudes and beliefs about children in general. What is your level of knowledge in the area of child development? Do you perceive children to have rights? How do you define discipline and what methods do you support? What is your attitude toward spanking or corporal punishment? What were the rules that pertained to children in your family of origin? Have you explored the dynamics of your family of origin? Have you allowed your true feelings to emerge about the ways adults interacted with you as a child? What did you discover that might influence your present attitudes toward children? Describe the distribution of power and control in your family and the positive or negative effects on family members. If you are able and willing to examine frankly the experiences you had as a child in your family of origin, you will be able to identify beliefs and feelings that undoubtedly will profoundly affect your professional activities with families and children.

Encountering Children in Placement

You'll be presented with many fieldwork opportunities to work with children in residential settings outside their family homes. Seeing children endure the separations and losses inherent in the foster care system (or that occur while children await adoption) can be quite painful. It is important to develop the ability to view these experiences through the eyes of the individuals involved. To be truly helpful, you must have the capacity for empathy even in situations that may be quite foreign to you. For example, if you have had the benefits of a loving home and family, it may be nearly impossible for you to understand parents who may seem unwilling or are unable to care responsibly for their children. You may be tempted to judge these individuals harshly or interact with them in an insensitive or angry manner. Why, you may ask, would this parent not do all that is required for the welfare of the child?

If you ask the question out of a sincere desire to gain more information and understanding of the situation, then it will be valuable to proceed in that direction. As you explore the various factors, internal and external, that are contributing to the dilemma, you may develop more compassion toward those involved. If, however, you ask the question in order to criticize or place blame, you probably will succeed in widening the gap between yourself and the parents. Although you may take some immediate comfort in preserving the notion of yourself as different or separate from those who experience certain difficulties in their lives, in no way will your actions encourage a helping relationship to grow; in fact, the potential for positive change will certainly diminish.

Empathy is a vital element in all human services work. To foster our capacity for relating to others from a position of empathy, we must learn to tolerate the ambiguities of human nature and be willing to "step into" the experiences of others. Issues surrounding the health, welfare, and protection of children test our strength by touching us at our emotional core. By refraining from making judgments that generally are motivated by fear, we can remain accessible to those we seek to assist. Regular discussion and consultation with others also will help us to maintain a healthy and objective perspective.

HOMELESSNESS AND POVERTY

One need only scan the local newspapers or listen to the daily news reports to be aware of the vast numbers of people who suffer extreme conditions of poverty here and abroad. The effects of deprivation in such basic areas as nutrition, health care, educational opportunities, and housing are far-reaching. Generations of economic struggle produce children who are ill-prepared to improve their plight and so the cycle continues. If your life has not been touched by this issue, you may find it hard to comprehend the mentality of scarcity that can often proliferate in an atmosphere of need and poverty. Children grow up with severely limited expectations of life in general and of their future in particular. People who serve as role models of economic or educational success may be too far-removed from the child's immediate environment to have any real influence. Dreams are discouraged in deference to a need to attend to basic survival. Aspirations that may seem average to those who enjoy economic security seem lofty and unrealistic to families who have known only a meager existence. A family in poverty that seems unsupportive of an ambitious child may only be trying to protect the child from a world that has held many disappointments.

Coping with Harsh Realities

Statistics that cite the growing numbers of homeless in our country are staggering, particularly when we remember that the United States is considered a wealthy nation. Fieldwork students, who are required to devote a certain number of hours to community services, commonly interact with homeless individuals and families in placements such as shelters, public assistance offices, and outreach programs. Homelessness is a difficult problem to grasp; our natural inclination is to protect ourselves from the anxiety that comes from even imagining ourselves attempting to survive without shelter. Having a permanent residence is considered basic to the average life circumstance in our society and is a symbol of security and belonging. To lose one's home is to forfeit protection and safety.

Your reactions to this issue and to the people whose lives are touched by it may range from sadness and depression to fear and avoidance. The enormity of this difficulty as it comes to life in a family with whom you are working probably will feel overwhelming at times. The contribution you are able to make may seem trivial or perhaps meaningless. Often there will be nothing you can say that seems to adequately address any of the family's concerns. A homeless person may express anger about his situation or point out that you couldn't possibly understand what he is going through. The fear a homeless mother has for the safety of her children may haunt you at the end of the day.

The reactions that you have may be difficult to manage at times. It's extremely important that you develop skills that will allow you to achieve some measure of separation between your fieldwork and your personal and family life. If you can develop a tolerance for powerful emotional states in yourself, you will be able to invite and accept these states in others.

VIOLENCE

We all agree that we live in a violent society. Every day, all manner of violent incidents occur in urban and rural areas alike. We all have become cognizant of the need to take measures to ensure our own protection and that of our possessions. Despite the prevalence of violence, we function on a daily basis by engaging in a level of denial that allows us to perceive the world as a predominantly safe place. It can be a shock, therefore, when we are confronted with the effects of a truly violent action.

Students in fieldwork frequently opt to work with segments of communities or populations for whom violence is a regular feature of daily life. Violence is an issue in outreach programs for juvenile runaways and prostitutes, diversion programs for at-risk youth, detention facilities, and mental health centers for families dealing with domestic violence and child abuse. Students also face issues of violence in crisis situations involving rape victims or people who become the target of criminal activities motivated by hatred or intolerance of racial, ethnic, or religious differences.

Another way students encounter violence is in their work with people who engage in self-destructive acts such as suicide attempts, self-mutilations, and serious eating disorders. It is unlikely that one can have a series of fieldwork experiences or a career in the human services without confronting individuals who are struggling with the mental health problems that sometimes give rise to these events. Such individuals arouse potent feelings of anger, sadness, fear, and helplessness on the part of human services workers. Faced with the uncertainty of a client's very survival, it can be quite difficult to realistically assess your degree of responsibility in the outcome of the situation. How can we attain that delicate balance between too much and too little objectivity? What are the limits of our responsibilities when a client's life is at stake? Experienced workers often consider the following general guidelines:

1. Remember to distinguish your feelings from those of your clients.
2. Understand that a client who feels overwhelmed or helpless may generate a similar emotional response in you.
3. Maintain appropriate boundaries between yourself and your client. For example, avoid the temptation to bring a client's problems home with you.
4. Be aware of any rescue fantasies that you may have and seek consultation when they occur.
5. View yourself as part of a treatment team or as one branch of a system of resources available to a client.
6. Consider other supportive resources for your client, including all appropriate family or community resources.
7. Provide yourself with adequate social support.

Managing Your Emotions

The most typical response to situations involving violence is fear. The intensity of this emotion will vary among individuals in any given circumstance, but some components will be universally experienced and are rooted in our genetic and evolutionary backgrounds. Take a moment to review past occasions when you have experienced strong feelings of fear or anxiety. What physiological symptoms became noticeable? Was your ability to think affected? Was your capacity for making sound judgments impaired in any way?

Even when we listen to descriptions of violence, we experience some of the same reactions as those we would have if we witnessed a violent event. Fieldwork students and experienced professionals who work with domestic violence, rape, child abuse, gang violence, and criminal populations often develop responses not unlike those of trauma victims, including feelings of generalized anxiety, a period of heightened awareness of their surroundings (usually referred to as hypervigilance), and a preoccupation with thoughts or dreams related to their experience. If we can learn to be aware of how a client's description of a violent event is affecting us, we can turn to supervisors and colleagues for appropriate debriefing and consultation.

When working with scenarios involving violence and trauma, there probably is no way that you can remain entirely unaffected unless you remove yourself entirely from the situation. At times and for certain individuals, this is the correct course of action. In any case, it generally is wise to balance your caseload or fieldwork activities so that you are not immersed in a sea of cases and interactions that involve violence and trauma. For example, if you are providing counseling in a school setting, request that your cases reflect a diversity of problem areas. Talk to your supervisor about the nature of your caseload and ask for support. Discuss any concerns you may have about managing your own stress in fieldwork seminar meetings; other students will have similar concerns and may make valuable suggestions regarding stress management.

Finally, it is inevitable that situations with an element of violence will provoke feelings of helplessness and loss of control. For human services workers who pride themselves on their ability to find meaningful ways to help others and influence events, these feelings can be a source of great concern and discomfort. Experience teaches us to be realistic in our objectives, to value the extent of our contributions and to learn to tolerate our limitations when dealing with enormously complicated issues. Recognizing that we have limitations in no way diminishes what we are able to give.

Chemical Dependency and Substance Abuse

Chemical dependency and substance abuse is one of the most challenging and frustrating problem areas encountered in the human services. As with other problem areas, there is no single set of rules or approaches that, once mastered, will allow the worker to feel totally at ease. Complexity is a key aspect of this area of practice. That being said, it also is true that our body of knowledge in this field has grown substantially over the years and we now are better able to grasp the bio-psycho-social interplay of factors than ever before. Chemically dependent individuals can achieve relief from their addictions assuming they are willing and able to undertake the lifelong process of recovery. Because it typically is believed that recovery is a process rather than a discrete event, individuals who seek help are asked to make a substantial commitment to their own treatment. This commitment is ongoing and can be very difficult, both for the person in recovery as well as for their friends and family. The recovery process is typified by incremental successes interspersed with relapses; this pattern requires that the client, family, and treatment team sustain their support during times of considerable frustration and uncertainty.

Physical and Psychological Dependence versus Abuse

Whether or not you have substantial expertise in the area of substance abuse and chemical dependency, you can anticipate encountering individuals with substance abuse problems. Drug and alcohol abuse occurs with great frequency in all social and economic strata, and all human services workers must develop a basic familiarity with the signs and symptoms of substance abuse. There is much to learn, and it will be to your advantage to pursue coursework and reading beyond this introductory discussion. To begin, it is important to note the distinctions that exist between *abuse,* which is the excessive and inappropriate use of alcohol or drugs, and *dependence,* which can be either psychological or physical. *Psychological dependence* is often reflected in an inability to refrain from the use of a substance due to the presence of cravings and a preoccupation with the desire for continued use of that substance. This desire may exist in spite of evidence that continued use is a threat to the person's well-being. *Physical dependence* is characterized by the presence of physical withdrawal symptoms that arise when use of the substance is discontinued. Both psychological and physical dependence may generate strong feelings of a loss of control over one's behavior. Psychological and physical dependence may exist simultaneously, in which case the user may become overwhelmed by feelings of being trapped and helpless. Although psychological dependence may exert a powerful influence on the day-to-day thoughts and feelings of the user, physical dependence poses a threat of medical emergency, if withdrawal is initiated without appropriate medical supervision. We encourage you to learn about patterns of drug and alcohol use in your area or client population. A particular neighborhood, community, school site, or even countywide area may experience problems relating to substances that have become prevalent during a particular period of time. Your awareness of this information may prove valuable in expediting an appropriate assessment and referral.

Family and Social Dynamics

Chemically dependent individuals generally enter into extremely dysfunctional relationships with family members and friends. Denial of the severity of the problems by the user and his or her significant others is common. When those who are involved with the user enter into behavior patterns that tend to support the drug or alcohol abuse, there may be a system of *codependence* present. Consequently, effective treatment usually entails ongoing family involvement. In order to maximize the possibility of treatment success, both the individual and those who surround that person must be willing to engage in an examination of reciprocal behaviors that may have been established over a lengthy period. It is not unusual to find that a substance user and his family members were exposed to relationships with other chemically dependent family members at an earlier time. Frustration and emotional upset should be expected in clients and family members who enter the recovery process.

Anyone who has worked in this field is aware of the need to develop a sophisticated and realistic posture when working with substance abusers. For example, although you might want to believe that most or all clients would not intentionally lie or attempt to deceive you, that is not the case here. Drug- and alcohol-dependent persons are not inherently bad or predisposed to any of these behaviors, yet they often have established a pattern of secrecy and denial in order to avoid detection and maintain their habit. Over time, the boundaries that separate inappropriate from appropriate behavior blur and patterns of manipulation and deceit become the norm. As you might expect, trust is eroded in such a structure, and feelings of anger and loss of control emerge in all parties involved. Family members often wonder why the individual simply does not stop the destructive use of the substance when it is clear that continued use is damaging the person's health and important relationships. They think: Doesn't he/she love me enough to stop? How could he/she lie to me about wanting to stop using? Why don't I stop supporting him/her entirely? Why do I feel so out of control myself? The worker will strive to educate clients about these patterns, confront their codependent behaviors, and provide support to them over the course of recovery.

It is imperative that family members be referred to community programs that are designed for these specific problems, such as Al-Anon and other relevant support services.

VIGNETTE ONE

Margaret is a 63-year-old diabetic with a known alcohol dependence dating back well over 20 years. She went through inpatient rehabilitation 10 years ago and has attended Alcoholics Anonymous meetings sporadically ever since. Her husband suspects that in addition to alcohol she may be abusing a variety of pain medications that she received from her physician following back surgery 2 years ago. He noticed recently that she also has prescription medications from other doctors, but he has avoided asking her what they are for because she ordinarily is defensive about such questions.

You are on call at a local community mental health clinic. Margaret's husband calls to ask for advice about what to do. It seems that his wife has begun drinking more and more lately, and he is concerned about a worsening of her diabetes. When you suggest that Margaret be evaluated by a physician he sounds irritated and says that in his opinion physicians may be part of her problem. He asks whether there is any way that you could call her, without referring to his suspicions about her prescription medications. He also asks that you avoid any mention of hospitals because Margaret is suspicious of them due to her past experience.

QUESTIONS FOR DISCUSSION

1. What can you identify as Margaret's needs?
2. What steps would you recommend for Margaret, in order of priority?
3. What are the needs of her husband?
4. What steps would you recommend for Margaret's husband, in order of priority?
5. What are some possible resources in your community for Margaret? for her husband?
6. How do you imagine Margaret to be feeling?
7. How do you imagine her husband to be feeling?
8. What additional questions would you like to ask Margaret's husband to assist in your assessment?
9. Would you consider calling Margaret? What would you say to her? Would you honor her husband's requests not to mention certain topics? Please explain your thinking.
10. Have you had any experience with someone with problems similar to Margaret's? What was the experience like? Did you try to help?

ANALYSIS

Assessing treatment needs for someone like Margaret or her husband is never simple but does become easier as your experience grows. In this circumstance, one issue facing the worker is the level of denial and resistance that is present. Here the worker is called upon to do more than express empathy. He or she must be appropriately confrontive as well, and impart information and make recommendations in a clear and confident manner. Of course, change may be hard to come by. Clients don't always agree with their workers and they may fear that negative consequences will ensue if they allow themselves to take frightening steps. Mistrust and denial can thwart the best efforts of a well-intentioned worker. As always, it's important to do the very best job you can and remember that there are reasonable limits to what you can control. It's crucial to keep some distance in order to maintain a clear, realistic perspective.

Death and Dying

We operate much of the time in a state of denial about the inevitability of death. Given the sadness and fear associated with death, it's not surprising that we regard it as a topic to be avoided. In spite of our discomfort, we all have an eventual expectation of the loss of our friends and loved ones and of our own death. Because we assign the greatest possible value to human life, it is understandable that we fear any process that culminates in the end of life.

Depending on your age, you may or may not have experienced the loss of someone close to you. If you have, you know that it is normal and natural to feel great sadness and longing for someone who has died. The intensity of these feelings usually relates to the position of importance that the person held in our life; many of us struggle to deal with painful losses for a number of years. Whether continued life experiences lead to a greater acceptance of death varies from person to person. People differ enormously in their response to death.

We cannot expect our clients to be uniform in their response to death and dying. Our religious and philosophical beliefs influence the manner in which we regard death; some individuals rely heavily on these beliefs, while others need help managing the powerful feelings that emerge. You will need to be emotionally present for clients, to express genuine empathy for their experience and offer constant support. Perhaps now more than any other time, you may be tempted to impose your own beliefs and perspective on a client. You must strike a balance between offering support and sharing your views, and recognizing a client's right to his or her own beliefs.

There are many ways we can become less helpful to clients in their time of need. Emotions we felt during prior losses will resurface whenever we are reminded of those persons and events. When this occurs, we may hope to act quickly in order to end a client's obvious pain. In doing so, we will interfere with the natural progression of the client's unique grieving process. We become less available to the client as a resource to rely on during a time of need.

One dynamic that is important to note here is the relatedness of all losses. Whenever one loses an attachment to another of any kind, it is a type of loss. A separation from a spouse is a loss, as is the death of a child. Although one event may not equal another in terms of intensity or pain, both have the potential to remind us of prior losses. In this manner, our memories of loss may overlap and influence our experience of subsequent losses. Workers must consider this when they evaluate an individual client's response to a given loss. A client may need to resolve feelings about earlier deaths or losses in order to allow for the eventual acceptance of the present scenario. Depending on the complexity of these variables, the amount of time necessary for resolution and the degree of support required will vary. It is always important to examine whether an individual has an adequate support system and to assist in expanding his or her connections to others whenever possible.

In order to clarify your own approach to death and loss, answer the following questions:

1. What are your beliefs about the process of death? Does this include a belief in an afterlife?
2. How might your beliefs affect your work with clients? What if you had a client who was diagnosed with a terminal illness?
3. What feelings do you experience when a client describes his or her reactions toward death and loss? Describe how your feelings are an outgrowth of your life experiences.
4. How would you describe your capacity to cope with the loss or death of a loved one?

VIGNETTE TWO

You are a student in a metropolitan hospital associated with the department of social services. Among other duties, you are responsible for providing counseling and referral services to a variety of medical patients and their families. One of your clients is a 65-year-old man whom you have known for a few weeks and like and admire. As usual, you stop by to see him one afternoon and he casually discloses that he has just received a diagnosis of inoperable cancer. He seems resigned to his fate and is somewhat quiet and withdrawn, which is quite a change from his usual demeanor. He startles you by asking what you think about his diagnosis and his chance for survival.

QUESTIONS FOR DISCUSSION

1. How do you imagine yourself responding?
2. What do you think this man's thoughts and concerns are likely to be?
3. What might be an appropriate response to him in this scenario?
4. What would be the most difficult aspect of the situation for you? for him?

VIGNETTE THREE

You are a student in a metropolitan hospital associated with the department of social services. You are primarily responsible for working with preadolescent children and their families. Currently, your client is a 12-year-old boy with a virulent form of brain cancer and an uncertain prognosis. The physician in charge of his care has reported that the boy has become resistant to accepting his chemotherapy—he recently crawled under a table and refused to come out rather than submit to the unpleasant procedure. The physician has asked you to meet with the boy to try and help.

QUESTIONS FOR DISCUSSION

1. How you would you feel about working with this boy?
2. How might your feelings interfere with your effectiveness?
3. How would you respond if the boy remarked that he had little reason to accept the continuing treatment?
4. What would you say if the boy asked you what you would do in his place?
5. How might you explore with him the feelings he has about the possibility of dying? What would you do if he resisted your attempts to converse on the subject? Would you persist in your efforts?
6. How would this scenario be different if the patient were an adult?

ANALYSIS

Professionals who practice in medical settings will face issues of physical loss and death more often than their counterparts in other settings. As a result, these workers must become accustomed to having open, frank dialogues with clients who are facing extremely difficult scenarios.

As you might imagine, seasoned health care professionals develop a mature approach that is characterized by clarity and the expression of genuine empathy. Clients appreciate the availability of a worker who is not intimidated or overwhelmed by the topic of their illness. In some circumstances, a client may be reluctant to discuss either the nature of his illness or his fantasies about what the future may hold. At these times, the worker may provide an example of open expression to the client. It should be remembered, however, that workers must always begin where the client is, and adjust our approach accordingly. Clients who are fearful (or even terrified) of their illness or prognosis may need a longer period of time to discuss their feelings or require a gentler approach than those who are better able to accept harsh circumstances.

ACQUIRED IMMUNE DEFICIENCY SYNDROME (AIDS)

Although medical breakthroughs in the treatment of AIDS are frequently reported, it is clear that AIDS and AIDS-related syndromes are among the most feared disorders of our time. The term *AIDS,* like the word *cancer,* raises our level of anxiety as few others can. The mere possibility of contracting this disease is terrifying to many of us. Still, the prevalence of AIDS and its rate of growth make the likelihood of our contact with individuals who have AIDS or are HIV-positive more and more certain.

Imagine contracting an illness that caused you to become a social outcast at the same time that you were facing your greatest fear and had the strongest possible need for support. How might you respond? You don't have to be a therapist to comprehend that a serious illness such as AIDS calls for a close involvement on the part of loved ones and friends. Yet we are all familiar with how our society seeks to isolate those who are the most seriously ill—mostly for our perceived benefit rather than theirs. It is certainly understandable that an illness as frightening as AIDS may tempt us to avoid contact with patients who have contracted the disease. That posture may work until someone we are close to, or even we, become ill. At that point the illness is no longer an abstraction—it is real, and it no longer can be ignored. It is worth considering not only how you might respond to an AIDS patient who is your client but also how you would feel if you had the disease yourself. What would you want and need the most?

One thing that's important to consider is the impact of a major illness such as AIDS on an individual's overall identity. Previous perceptions that a patient may have had of himself may be displaced by the presence of the illness. No longer is he a school teacher, engineer, administrator—he is a person with AIDS. As a human services worker, you must remember that a person is not merely a reflection of his or her circumstance—a person is a complex human being who is defined by many factors other than serious illness. Can you think of any time that a single event or process significantly affected how you felt about yourself as a person?

Those in the helping professions have developed a range of services for AIDS patients. Groups for patients and family members help to combat feelings of isolation and provide useful information. Interaction among families and AIDS patients in a multifamily setting allows for the sharing of information across family lines and helps members to realize their essential similarity and connection with others. Home-based outreach is provided for seriously ill AIDS patients who are unable to travel to services.

As with patients who have other life-threatening illnesses, the possibility of death becomes an issue for some AIDS patients and their families. In the event that the illness threatens death, the worker will be called upon to speak openly with the patient and his or her support network. Remember, clients often will follow our lead in entering into a discussion of difficult topics such as death. Unless we are willing and able to share some personal reactions and ask difficult questions, our clients may be deprived of the opportunity to express their feelings.

The following questions may help you clarify your own feelings about AIDS:

1. Would you be willing to work with AIDS patients? Why or why not?
2. How would you respond to the knowledge that a family member of yours had contracted AIDS?
3. What ideas do you have for enhancing your community's response to AIDS?
4. Would you respond to a child with AIDS differently than you would to an adult with AIDS? In what manner? Does means of contraction make a difference in your viewpoint? Explain your reaction.

Suggested Readings

Briere, J. (1992). *Child abuse trauma.* London: Sage.

Gelles, R. J., & Loseke, D. R. (Eds.). (1993). *Current controversies on family violence.* Newbury Park, CA: Sage.

Gordon, J., & Shontz, F. C. (1990). Living with the AIDS virus: A representative case. *Journal of Counseling and Development, 68*(3), 287–292.

Herman, J. (1992). *Trauma and recovery.* New York: Basic Books.

Hernandez, D. (1993). *America's children: Resources from family, government and the economy.* New York: Russell Sage Foundation.

Hoffman, M. A. (1991a). Counseling the HIV-infected client: A psychosocial model for assessment and intervention. *The Counseling Psychologist, 19*(4), 467–542.

Kottler, J. (1992). *Compassionate therapy: Working with difficult clients.* San Francisco: Jossey-Bass.

Peters, R., McMahon, R., & Quinsey, V. (Eds.). (1992). *Aggression and violence throughout the lifespan.* Newbury Park, CA: Sage.

Roth, L. (1987). *Clinical treatment of the violent person.* New York: Guilford.

Chapter Eight

Experiencing Closure

- Endings: Patterns and Dynamics
- Assessing Your Fieldwork Experience
- Your Performance Evaluation
- The Process of Separation
- Suggested Readings

ENDINGS: PATTERNS AND DYNAMICS

Just as we experience the dynamics of beginnings in uniquely personal ways, the manner in which we manage endings also reflects our own history and inclinations. A few moments of thought about the endings you already have moved through in your life usually will reflect a pattern of behaviors, thoughts, and feelings that typify *your* style of coping. You may notice a general tendency to avoid people or even tasks and responsibilities as the final days of a commitment or relationship approach. You may find yourself looking forward to the end and feeling relieved that the experience will soon be over. Perhaps you delay giving any thought to the fact that an ending is near, or think about it only in a very general way, without attending to the implications and consequences. You may even feel a deepening of your connection or sense of loyalty, which prompts you to find a way of continuing in a setting, perhaps by staying on in some other capacity. Most of us confront endings with some degree of ambivalence. We all on occasion have sidestepped what we anticipate will be a difficult closure by quietly and unobtrusively drifting away. How endings affect us depends on how attached we have become to people, responsibilities, and roles. Our interpretations of past separations and losses form the filter through which we view similar events in the present. As with all of the topics we address in this book, we again must emphasize the importance of developing a keen self-awareness about these issues. Knowing yourself well promotes a conscious appreciation of the subtle dynamics that constitute the process of termination. Having chosen the field of human services, you often will be called on to model appropriate interpersonal behavior; learning to say good-bye is one challenging aspect of this role.

Viewing Termination as a Process

In view of the complex nature of separation and loss, it helps to regard termination not as a discrete event that occurs at a certain moment of your final day in a fieldwork placement, but as an ongoing *process*. That final day is merely the culmination of a process that was set in motion at the onset of your fieldwork assignment. Unlike many other relationships or employment situations, in fieldwork specific parameters are agreed upon in advance by the student and the agency. With the average length of commitment varying from one semester to one calendar year, keeping the prospect of a known termination in the forefront of your thinking is both necessary and responsible. You hardly can expect to prepare co-workers and clients for a loss without being willing to consider your own experience in a thoughtful manner.

Having a finite period of time in which to accomplish a goal can be quite motivating. It is common to hear people describe how they are more productive when they have a set deadline than they are when tasks can be scheduled at their leisure. Knowing that your fieldwork time in a given setting is limited prompts you to project realistic and well-defined objectives. Although you must remain flexible enough to work within the constraints of each fieldwork environment, you can optimize your stay by prioritizing the opportunities to learn new skills and enhance existing ones. Being mindful of your timeline will stimulate you to request early and active involvement and will discourage you from taking a passive or inactive role at the start. When you strive to make the most of every moment, you communicate through your attitude how seriously you regard your training and practice. Supervisors and other co-workers take notice of this and usually respond by making the effort to ensure a high-quality training experience for you.

The termination phase of a human services internship consists of three major parts: (a) your own assessment of the total fieldwork experience, (b) evaluation of your performance by a supervisor, and (c) separation from clients, staff members, and the agency itself. In order to realize a fully satisfying termination, several concerns need to be considered. It is to these matters that we now turn our attention.

Assessing Your Fieldwork Experience

Although assessing the fieldwork setting and the quality of your overall experience should be an ongoing endeavor, students in their final weeks are more inclined to focus their attention in this direction. Academic semesters are coming to a close and students themselves are being assessed as they undergo final exams. Fieldwork supervisors are readying evaluations of the students who have been placed in their training facilities. Everyone is seeking to wrap up the work at hand so that they can begin planning for the next time period.

Some college and university programs require students to complete agency evaluations in written form. Students are asked to critique various aspects of the setting: its overall functioning, how well the client population is served, the structure of the agency, and its quality as a training site for interns. If your academic program does not require your participation in this process, it is worth giving thought to these and other issues because it will help you to be discerning when you are making decisions about future employment.

Assessment involves examining the various dimensions of a given situation, with the goal of understanding the particular contributions made by each element to the total experience. In the beginning of a fieldwork assignment, it is difficult to apprehend the whole or gestalt of what lies ahead; indeed, at the start of a new placement, many details must be rapidly assimilated, and it takes some time for an integrated picture to emerge. Immersed in our daily tasks and challenges, we often are only slightly aware that our focus is beginning to shift, that we are less preoccupied with fragmented details and more aware of the total picture. Gradually, we see the relation between components that may have seemed unrelated. As the novelty of our environment fades and unfamiliar tasks become more routine, our anxiety decreases and we regain our perspective. We begin to see the bigger picture and our place in it. Our recognition of these interrelationships continues to develop throughout our months of fieldwork; however, opportunities for reflection and examination are few and far between. It is only as we near the end of the semester or year and begin to separate from clients and duties that we find the time to engage in an insightful, intentional assessment of our experience.

Sharing Insights and Remaining Objective

As we've said, academic programs that require student participation in fieldwork routinely institute methods of assessing agencies and other settings that function as training sites. Although these approaches may differ in format, some form of regular communication between the fieldwork coordinator, faculty members, students, and agency supervisors usually occurs. Site visits are one method of ensuring a good fit between the requirements of the school program and the needs of the agencies. There may also be a paperwork component, which is comprised of forms designed to elicit descriptive information and satisfaction ratings from students and supervisors upon completion of the fieldwork assignment. The entire process aims to engender productive and mutually beneficial relationships among academic institutions and community service providers. Educational programs and agency training sites rely on each other to achieve their respective missions.

There are a number of distinct and significant areas to address when the time arrives to assess your fieldwork experience. Consider the following questions:

Questions for Assessing Your Fieldwork Experience

1. What were some of your initial expectations as the fieldwork experience began?
2. How specific were your goals and objectives? (If a contract was used, please refer to it now.) Did the way your objectives were worded affect whether you were able to achieve the particular objectives? How?
3. What additional information would have assisted you in developing suitable goals?
4. How satisfied were you with this fieldwork experience by the end of the first month? Be specific.

5. What concerns had developed by the end of the first several weeks?
6. In retrospect, what changes, if any, would you make in your level of activity during the first several weeks (that is, in pursuing opportunities, communicating your needs to your supervisor and co-workers, and so on)?
7. How prepared did you feel to assume the tasks you were asked to undertake?
8. What were your initial impressions of the administrative or managerial staff? your supervisor? other interns? How did these change over time?
9. Identify your particular strengths and weaknesses as they became evident in this setting and with this specific client population.
10. What steps did you need to take in order to develop rapport with staff members or clients in this setting? What was this process like and how successful were you?
11. Did you feel you received an adequate amount of feedback on your performance?
12. Were you able to ask for help or specific feedback when needed or desired?
13. What were the high points of your experience in this placement? Be specific.
14. What were the low points? Be specific.
15. Were you able to find a comfortable and effective place within the organizational structure of this agency? What factors helped or hindered you?
16. As a result of your experience in this placement, in what ways do you feel more prepared for work in your field? How has your knowledge base expanded?
17. Overall, what is your level of satisfaction with your experience in this placement? What suggestions, if any, would you make to improve the quality of this setting as a training environment for future fieldwork students?
18. In retrospect, what would you like to have known about this setting before you chose it as your fieldwork placement?
19. Would you recommend this placement to other fieldwork students? Why or why not?

In making a thorough evaluation, it is helpful to look at each distinct element of the setting (administration, staff members, quality of supervision, opportunities for learning, client population, mode of delivery of services, and so on) to avoid arriving at a vague or general feeling about your experience. Feelings tend to reflect our immediate reactions to particular events, interactions, successes or failures, and will naturally fluctuate over time. Although it is important to honor this aspect of our experience, we must take a more objective and inclusive approach to our final evaluation if we hope to be fair in our assessment. Otherwise, we might allow a particularly difficult or unsatisfying task or interpersonal interaction to color our perceptions of the overall experience. Challenge yourself to examine perceptions, events, and interactions with others with a discerning eye. Analyze your own part in any given circumstance before considering other elements of the situation. You may find that you allowed one aspect of your experience or one relationship within the agency to cast an unfortunate shadow over the entire experience. Remember, other students who ask your opinion of a setting will rely on your ability to give a well-balanced critique.

Your Performance Evaluation

Several weeks before the end of your fieldwork assignment, you undoubtedly will become aware that the time for an evaluation of your performance is near. Academic programs often establish a means for agency supervisors to compile a range of information that is used in determining the grade or credit earned for completion of fieldwork hours. Typically, there will be a written assessment, which in some cases is supplemented by telephone contact or a face-to-face meeting between the supervisor and the seminar instructor or fieldwork coordinator. We strongly recommend that you take the earliest opportunity to familiarize yourself with the specific evaluation process used by your program. Evaluations tend to occur at a time when even the most organized student may feel swamped by final examinations, term papers, and other tasks related to fieldwork termination. Preparing early for the evaluation process results in less frustration and more productivity when time is of the essence!

Preparing for Your Evaluation

One of the most important things you can do to prepare for your fieldwork evaluation is to explore your own mindset about assessment and appraisal. In fields related to human services, an emphasis is placed on personal growth. We encourage ourselves and our clients to be less dependent on external circumstances as measures of self-esteem. We strive to develop and enhance our ability to perceive the intrinsic value of people and to suspend judgments based solely on performance issues. Psychological approaches to understanding and analyzing human beings include all manner of ways in which to describe thoughts, feelings, and behaviors. Some views are reductionistic in nature while others discourage any temptation to test or quantify aspects of being human. Some stress the importance of subjective analysis while others align with objectivity and standard scientific methods. Often, systems of evaluation seem to complicate or even contradict our inherent philosophy or belief system about how to interact with people in meaningful ways. Think about the various ways in which aspects of your abilities and achievements are judged. Do you agree with the methods for grading academic performance, or for rating conduct and achievement, or do they seem too arbitrary, too subjective, or even artificial? Whether or not we wish to subscribe to these processes of ranking or classification, these systems are widely used to categorize and communicate about our abilities. It is worth the effort to understand your reactions to being evaluated and to come to terms with them in a way that prompts you to participate on your own behalf rather than engage in resistance or sabotage.

Some supervisors prefer to prepare their evaluation of your fieldwork performance and professional development without additional input. Others may rely, in part, on the observations of staff members and colleagues who have had the opportunity to work more closely with you on various tasks. Still others may ask you to duplicate the evaluation forms so that you can engage in a self-assessment, which then will be used for comparison and discussion. This type of process allows you to participate in an evaluation that is a cooperative effort. Be flexible and remain open to suggestions your supervisor may make regarding your participation in the evaluation. Academic institutions all have different methods of performance evaluations as well. Make sure you know whether you are required to complete any formal documents or paperwork as part of your own evaluation. Your supervisor will appreciate receiving any necessary paperwork well in advance as well as an occasional reminder as the due date approaches.

The Evaluation Meeting

Once evaluation forms are completed by your supervisor, the information will be shared with you in some manner. Having supervised students in several types of fieldwork settings, we have come to favor individual meetings that allow enough time for a lengthy discussion of all material included on the evaluation form. If your supervisor does not initiate a meeting for this purpose, we strongly encourage you to request one. Approach the meeting with an open mind and allow yourself to participate fully in the process. Ideally, the forms will have provided space for your supervisor to include descriptive commentary in addition to using a standard rating scale. If not, ask for clarification of the ratings so that the evaluation is more meaningful to you. If your supervisor explains how he or she arrived at a certain rating, you will become more informed about the different factors that were considered. Numeric ratings and terms like *satisfactory, deficient,* or *outstanding* are too vague to be of any real benefit.

Of course, it is not uncommon to find that you do not completely agree with everything on the evaluation. Your own self-knowledge, educational background, and previous professional endeavors allow you a perspective that naturally is quite different from that of a supervisor who has known you only in the context of the fieldwork placement. For example, you may rate yourself on the performance of a particular skill relative to your ability in the early stages of the internship. A supervisor, however, may also consider how your performance compares with other interns at the agency. Supervisors who are extremely busy may regard the average student's need for training as burdensome; their wish for a student to be more self-sufficient and independent may be reflected in their rating. Personality traits or interpersonal style of both the student and the supervisor can affect the entire process of evaluation as well as the final ratings. To the degree that you assert yourself by asking clarifying questions, you will increase your understanding of the evaluation and your supervisor's method of assessing you and your performance. Of course, supervisors vary as to how they welcome your input and queries. Be sure to phrase your questions or comments in a respectful rather than challenging manner. If you find it hard to accept criticism, or if evaluations create a level of anxiety that is difficult to manage, you may wish to ask for a copy of the completed forms in advance of the meeting. This way you can review the feedback, clarify your responses, formulate any questions or comments you may have, and gather your composure prior to the meeting. You will then be able to engage in a more productive and satisfying interaction.

Most students will experience some degree of anxiety about the evaluation process, especially if they have fully invested themselves in their fieldwork activities rather than viewing them simply as part of a requirement that must be fulfilled. In preparing for your evaluation, it may be helpful to consider the following questions, which will help you become more aware of your own reaction to being evaluated:

What have earlier evaluation experiences been like?
How were you affected by evaluations you have received?
What do you anticipate your responses will be in the present situation?
What does an evaluation at this time and by this supervisor mean to you?

The Process of Separation

One of the most complex aspects of human relationships is the process of separation. If you take a moment to contemplate all of the separations you have already experienced in your life, you will begin to notice the personal tendencies you exhibit when you are faced with a separation. Consider the many natural developmental occurrences that involve an element of separation: a child's first day of school; graduating from one class or school to another; relocating to a new geographical area; ending a friendship or romantic relationship; losing or giving up a job; divorce and death. Probably the most difficult part of dealing with separations is contending with the losses that are usually inherent in these experiences.

Experiencing Loss

Despite our repeated experiences with loss as a natural consequence of growth, development, and maturation, human beings often fear the inevitable discomfort of these events. Some losses, such as terminating at one employment setting to accept a position at another, contain the potential for both the sadness that accompanies leaving familiar and comfortable surroundings and the excitement that occurs with any new challenge. At times, we instigate the change and therefore experience the losses from a position of greater control. More often, we must come to terms with change that is uninvited and, not infrequently, unwelcome. In these circumstances, we struggle to manage a range of ambivalent feelings. If we haven't given proper attention to feelings of loss in the past, our subsequent responses may be even more intense. Can you recall a loss experience in your life that you avoided or inadequately mourned?

Grieving is a valued aspect of fully living our lives, although it often is regarded as unwanted or unpleasant. Our difficulties in tolerating our own feelings of grief and loss are evident in the attempts we make to stifle or avoid these same feelings in others. We may find it extremely awkward to be with persons who cry uncontrollably in their grief, especially if we tend to approach situations by attempting to mold, influence, or control others' reactions. We fear that grief will be overwhelming if just allowed to run its course and we try to protect ourselves from the pain and hurt attached to the losses in our own lives. The longer our lifespan, the more separation and loss we will experience; thus, learning to say good-bye and to tolerate our emotions is a valuable life skill.

Expressing Loss

When dealing with a separation or loss, it is important to allow yourself to grieve in the ways that are right for *you*. No two people say good-bye or mourn a loss in the same manner, and we must be careful not to impose our style of grieving on others. Cultural conditioning, religious beliefs, family patterns, and the specific circumstances of any loss all will contribute to the manner in which an individual grieves. Much has been written about how human beings process a loss experience; several emotional aspects appear to be common. They include feelings of anger, disbelief, denial, depression, and, finally, acceptance. Although the literature on death and dying suggests that specific stages occur along the path to acceptance, individuals do not move through these steps in a well-defined, linear progression. However, being given sufficient opportunity to express various emotional states appears to provide great benefit to persons who experience loss.

As a human services student, you will serve as a model to others in many ways. Your ability to identify, tolerate, manage, and express your feelings about separation and loss experiences will help you sustain your career and will influence others with whom you interact. Some students already are equipped to care for themselves and others during a time of loss, having learned adequate coping skills in their families. Others will need to discover the benefits of putting difficult experiences and feeling into words and sharing them with others. Talk with others about the ways in which you actively grieve in times of loss. Create rituals for yourself that help you to express your thoughts and feelings. (You may wish to draw on meaningful aspects of your cultural, religious, and spiritual background.) Learn to utilize the support of others through group experiences. Developing a comfort level with any mode of expression requires practice.

Terminating with Clients

Knowing what you do about the process of separation, it should be obvious that terminating with clients in a fieldwork setting requires both personal awareness and considerable skill in facilitating and coping with endings. Termination must be tailored to each situation, depending on the type of relationship you have with a given client. The nature of the work that defined the focus of the relationship, as well as its duration and intensity, must be considered. Most studies concerning the tenets of an appropriate and mutually beneficial termination focus on the relationship between counselor and client. Indeed, much of the literature on this subject speaks to this specific aspect of the helping professions. However, a satisfactory termination is an important feature of the many relationships inherent in the fields of social work, health, education, public administration, human services, and psychology.

Fieldwork students with whom we have worked have voiced the following concerns about facilitating terminations with clients:

"How early should I begin talking about termination with my clients?"
"How do I encourage my clients to discuss the feelings they may have about the fact that I am leaving?"
"How do I handle my client's silence when the subject of termination arises?"
"I'm afraid my clients will be angry with me."
"I'm afraid my clients won't trust me if they know I'll only be here for a limited time."
"What special considerations are there in terminating with child clients?"
"I'm afraid my clients will be disappointed in me."
"How can I leave if I haven't achieved my goals?"

All of these concerns will arise in the process of bringing closure to a fieldwork experience. As mentioned earlier, advanced knowledge of the approximate length of a fieldwork placement enables students to prepare clients early on for their inevitable departure. The concept of informed consent depends on the notion of disclosing all relevant information. Without dwelling on this information in a prolonged manner, you can tell clients in the early stages of the relationship that you are a student in training and that your stay in any given facility is limited. Some clients may hesitate to invest fully in a relationship that will exist for a shorter time than might be preferred. Clients have the right to determine to what degree they will enter into a helping relationship and what information they will reveal (and in what manner) based on full disclosure of the parameters of the relationship.

As we learn to attend to our own feelings in situations of loss and separation, we often increase our ability to invite, listen to, and talk freely about the emotional responses of clients. A client may experience multiple reactions when the subject of termination arises. It is understandable for them to feel disappointed, angry, depressed, sad, hurt, tolerant, forgiving, selfish, dependent, fearful, hostile, or even betrayed when facing the loss of a relationship that they have experienced as helpful. Your role during this time is to invite, encourage, tolerate, and discuss any feelings that your clients may have. Honoring their feelings (rather than resisting or disregarding them) assists clients in moving through the different phases of saying good-bye. If we allow ourselves to be wounded by clients' expressions of anger or disappointment, we will hinder their attempts to work through their experience. We should avoid sending subtle messages to clients that deter or impede further discussion (for example, by changing the subject, failing to address concerns directly, introducing humor to dilute the intensity of the client's feelings, and so on). Remember that just as you have a right to move on at the end of a fieldwork commitment, clients have a right to fully express their experience of the termination. If you encourage clients to process all the feelings attached to terminating the relationship, you will do a great service for those who find themselves approaching another helping professional in the future.

Child clients do require additional time to adjust to the loss of a relationship. Students working with children in the context of a helping relationship need to consider the child's ability to understand endings, the coping skills that the child possesses, and the child's need for extra support in progressing through the termination process. Age, maturational factors, cognitive development, and emotional makeup all shape the experience of endings for children. We look through the child's eyes to realize that commonplace concepts such as time have distinctly different meanings for children. It is not enough to use words or phrases like *soon* or *in a while* to describe an imminent ending; you must explain the time frame in a visual and concrete manner. (For example, toddlers and preschool-age children can participate in coloring in or crossing out the boxes on a calendar as a way to mark time.) It can be especially challenging for students to bring their helping relationships with children to a close; they may feel as though they are "abandoning a child in need." Although nothing can completely alleviate this feeling, it may help to focus on the value of the relationship the worker and child were able to develop, however brief or tumultuous it may have been.

Finally, one of the more common frustrations for people in the helping professions is the lack of any knowledge of the concrete results of their work. Sometimes this is because endings occur before projects have been completed or objectives have been fully met. In such situations it's difficult to gain a sense of accomplishment, particularly when outcomes are difficult or impossible to quantify. Students who are in the process of terminating their fieldwork assignments often express frustration at having to leave before they have achieved a sense of completion. How do we know for certain that we have been helpful in the way we intended? How can we measure the degree to which our efforts have been successful or effective when it may take years for the results to become evident? These can be troublesome issues for students and professionals alike. All human services workers must learn to find gratification in incremental changes. Measuring our efforts in terms of the internal values and the beliefs that prompted us to pursue a career in the helping professions is likely to be more sustaining than looking for external validation. Helping means taking full responsibility for the small but important role we play in an outcome we can neither predict nor control.

Separating from Co-Workers and Colleagues

The final step in bringing closure to your fieldwork experience will be that of saying good-bye to your supervisor and to staff members and other interns who have been your colleagues over the past months. At this juncture students often become intensely aware of the investment they have or have not made in relationships at the fieldwork site. It can be quite revealing to discover that the impending separation evokes strong feelings of sadness and loss or, to the contrary, that it has relatively little meaning for you. Whatever the case, the opportunity for learning about yourself and the way you relate to others can be quite valuable. The primary component in the work we do is our capacity to establish fulfilling and effective interpersonal relationships. Each step we take along the path of professional development signals whether or not we are continuing to enhance our interpersonal skills.

Relationships formed in fieldwork can be among the most satisfying of your academic years and may serve as important contacts for networking in the future. Advanced degree programs all require letters of recommendation from academic faculty and agency professionals who can vouch for your ability to succeed. It is difficult to approach a former supervisor or instructor to ask for such a letter when you have not established a relationship. The more effort you put into cultivating such relationships early on, the better your chances of drawing on them for support as you progress in your endeavors.

Many times in this text we have emphasized that the learning that takes place in fieldwork settings occurs in many reciprocal partnerships. The roles of student and teacher are always in flux, changing and adjusting as both parties become involved in various tasks, projects, and interactions. As fieldwork supervisors and university instructors, we are constantly learning from the students, colleagues, and mentors with whom we interact. Recognizing this makes daily interactions potentially more exciting. It also fosters and reinforces an atmosphere of individual respect wherein human beings can be valued without the constraints of preconceptions and stereotypes.

As you near the end of your commitment to a particular fieldwork setting, take time to decide with whom you want to have closure and how you wish to approach it. Allow yourself to prepare for saying good-bye; consider what specific things you want to say or do to facilitate your departure for yourself and others. There will probably be certain staff members, interns, or a supervisor with whom you have developed a special rapport. You may want to express appreciation to particular people who gave you support when you needed it the most. Perhaps you learned some valuable things from a casual conversation with another fieldwork student, and it seems important to let her know. You can avoid the disappointing experience of feeling rushed or missing an important contact by allowing yourself sufficient time and involving others in your planning. Terminations that are thoughtfully processed can be very gratifying.

Suggested Readings

Bowlby, J. (1988). *A secure base*. New York: Basic Books.

Collins, D., Thomlison, B., & Grinnell, R. M. (1992). *The social work practicum: A student guide.* Itasca, IL: F. E. Peacock.

Egan, G. (1990). *The skilled helper: A systematic approach to effective helping* (4th ed.). Pacific Grove, CA: Brooks/Cole.

Gutheil, I. A. (1993). Rituals and termination procedures. *Smith College Studies in Social Work, 63*(2), 163–176.

Kauff, P. F. (1977). The termination process: Its relationship to the separation-individuation phase of development. *The International Journal of Group Psychotherapy, 28*(3), 3–18.

Kubler-Ross, E. (1969). *On death and dying*. New York: Macmillan.

Martin, E. S., & Schurtman, R. (1985). Termination anxiety as it affects the therapist. *Psychotherapy, 22*, 92–96.

McRoy, R. G., Freeman, E. M., & Logan, S. (1986). Strategies for teaching students about termination. *The Clinical Supervisor, 4*(4), 45–56.

Chapter Nine

Planning for the Future

- Looking Ahead
- Anticipating Your Training Needs
- Strategies for Fieldwork Interviewing
- Pursuing an Advanced Degree
- Considering Professional Opportunities
- Suggested Readings

Looking Ahead

As you near completion of your fieldwork hours for this semester and are studying for final exams, trying to maintain some family and social life, and attempting to stay financially afloat, you may realize you need to give some thought to your immediate future! Any number of important decisions need to be made, depending on where you are in your educational and professional endeavors. If you are on track for completing a degree program, you may need to secure a new placement for the coming semester(s) in order to acquire more fieldwork hours. If you are graduating with your undergraduate degree, you may wish to seek entrance into a graduate school, find employment in your chosen field, or a combination of the two. In any case, it may seem as though the reward for all your hard work is more hard work! However, if a prolonged vacation or lengthy period of unemployment are not on your agenda, the time to explore your options is *now*.

Anticipating Your Training Needs

Let's assume that you are an upper-division undergraduate student who will be graduating in the near future. Your immediate concern is locating a new fieldwork placement so that you can complete the training hours required by your degree program. University fieldwork policies often require that you arrange interviews with potential agencies well in advance; if not, you must rely on your own initiative to get an early start. In this way, you will afford yourself the opportunity to be selective about the placements you explore and to choose (and be chosen by) the fieldwork setting you most prefer. Interviewing well in advance of your start date also allows you time to meet with your prospective supervisor and begin the process of designing reasonable objectives that foster your professional growth and are well suited to the practicum environment.

Locating an Appropriate Fieldwork Site

Academic programs that include a fieldwork component typically develop a roster of approved agencies—sites that provide high-quality training, sufficient opportunities for skill development, and adequate supervision. Whether you have specific goals in mind or need assistance in defining your areas of interest, we encourage you to seek advice from the fieldwork coordinator and from faculty members who teach fieldwork-related courses. Agencies vary widely as to client population, modes of service delivery, and range of available activities. It can be useful to consult with other fieldwork students who may have visited, interviewed in, or worked in a setting that you wish to learn more about. The impressions you glean from other students give you another perspective on life in that particular agency.

If you already have evaluated your present or previous fieldwork experience, you are well equipped to pose the questions and elicit the information needed for effective decision making. Review the process for formulating a thorough evaluation, and create a preliminary outline of your objectives for future placement.

Keep in mind the following points:

1. Ask yourself what you would like to learn and what type of setting might afford the greatest opportunity for this learning to occur.
2. What client populations have you yet to experience?
3. What might be an appropriate setting for building on your previous experience?
4. Are there social issues or mental health areas in which you would like to become involved?
5. What setting would allow you to exercise your strengths and address your weaknesses?
6. What information do you need about the setting in order to make the best decision? Be specific (times of available supervision, days and times of various activities, and so on).
7. What important concerns (if any) did you overlook in choosing your previous placement?
8. How will this fieldwork placement fit into your overall schedule?
9. Will experience in this fieldwork setting position you more advantageously for subsequent professional endeavors?

Strategies for Fieldwork Interviewing

Obtaining an Interview

Once you have determined which agencies hold the best potential for fulfilling your training and professional needs, it is time to schedule interviews. Again, honor the process by planning ahead and giving yourself sufficient time to prepare for each interview. Contact the agencies in which you are interested to confirm openings for new fieldwork students, and inquire as to any agency requirements regarding length of commitment. Ask any other pertinent questions that will help you decide whether you should proceed to the step of arranging an interview. If possible, request an appointment with the prospective supervisor or arrange for a telephone conference if busy schedules do not permit a face-to-face meeting. (The supervisor-student relationship is so crucial to the success of the placement experience that this step should never be overlooked.) Remind yourself to be courteous and professional from the first telephone contact on; first impressions are critical.

Managing Your Anxiety

Even under the most ideal conditions, an interview tends to be an anxiety-provoking experience. To counteract this effect, we encourage students to think of the interview as a two-way process: You are being interviewed, and *you are interviewing* a representative of this agency. You both are engaged in asking questions that will help determine whether a good match exists. Your role in this exploration is no less important or relevant than the other person's; subsequent to the interview, two decisions will be made. The agency may or may not offer you a fieldwork trainee position, and in response you may or may not opt to select this particular agency.

Looking at an interview from this perspective, it is easy to see that you are partly responsible for the interview content and direction. You cannot take full advantage of the allotted time if you are insufficiently prepared. To draft pertinent questions, use the guide you prepared in selecting the agencies you are approaching for interviews. Be clear and concise in your questioning and don't hesitate to bring a few notes into the interview room with you. Interviewers generally recognize these efforts as appropriate and will be impressed by your thoughtful preparation.

In addition to preparing your own questions, you should try to anticipate the potential questions you may be asked and allow yourself opportunities for rehearsal. Fieldwork seminars and assertiveness training groups provide ideal practice situations for this type of rehearsal. The feedback you receive from others can help you become aware of personal habits or tendencies that may contribute to the success or failure of an interview in subtle ways. Other students or group members can describe types of body language or speech patterns that may add to or detract from your efforts to communicate effectively. In the absence of these formal opportunities, trusted friends and family members can be enlisted to take part in rehearsal interviews. Walking through the complete interview in your mind (including a successful outcome), can be very useful; in fact, repetition can decrease anxiety noticeably.

Once you're in the interview, focus on your strengths. Some students and new professionals tend to deemphasize their abilities and focus on the areas in which they need further training and experience. Recognize your assets and be able to articulate them clearly. If you are asked about your special talents and skills, take the opportunity to demonstrate that you are capable and confident. If you also are asked to describe areas in which you may be deficient, be realistic about having much to learn but don't feel it's necessary to compile a lengthy list of particulars!

Finally, remember your appearance. How you present yourself says something about your ability to assume a professional role. Even if circumstances will require casual attire once you obtain a fieldwork position, an interview is a setting in which impressions are made on a number of levels. A good rule of thumb is to view the interview as an encounter between two professionals and to dress accordingly.

No matter how the interview goes, make sure you conclude by thanking the interviewer for his or her time. If you know that you are interested in the fieldwork placement, be sure to state this clearly, along with your ideas about how you might contribute to the mission of the agency. Don't leave without presenting a résumé that details your educational background and relevant professional experience. The interviewer knows that you are a student and your employment history will be brief; there's no reason to camouflage this fact.

It's a good idea to send a brief note to the interviewer a few days later, restating your interest in securing a fieldwork position and expressing that you enjoyed the interview. This might also be the time to mention any significant information that you may have omitted during the interview. In lieu of a note, a follow-up telephone call can be made.

Pursuing an Advanced Degree

If you are nearing the end of your current degree program and are considering further educational opportunities, you may be tempted to explore the world of graduate study. There are many diverse and exciting graduate school programs in all aspects of the helping professions and human services. Some are designed to lead to a master's degree while others progress toward a doctorate in a specific area. Decisions about graduate study require the consideration of several factors, including increasingly complicated licensing procedures (required for practice within several disciplines), an ever-changing employment picture, financial concerns and objectives, and the aim of achieving a well-balanced personal and professional life. Issues pertaining to your particular stage in life, your age, the status of your health, your role as a spouse and/or a parent, and other variables may determine the timing of pursuing an advanced degree.

Points to Consider

If further study is on the horizon, try to meet with faculty members you respect and admire who may have pursued a similar educational path. Many educators are happy to mentor or guide former and current students who are in the process of gathering information about relevant degree programs. There are many guides to graduate school education; they are readily available in libraries and bookstores and will help you become aware of the range of degree programs offered in your geographical area. If you are free to relocate, you can expand your search to include a broader range of possibilities. A short time spent perusing the pages of such a guide will soon result in a growing familiarity with basic elements, including entrance requirements, application procedures and deadlines, percentage of applicants accepted, tuition and financial aid packages, opportunities for part-time study, and the general ranking of the program compared to similar programs at other educational institutions.

Once you have identified more specifically the type of program you might wish to pursue, request information packages from several schools for comparison. Review the literature and formulate questions that might help you narrow the field. You may wish to arrange visits to particular schools to consult with an admissions representative or an advisor from the programs that interest you. Often, a school will make arrangements for current students or alumni of the program to talk with you about their experiences. These contacts provide a wealth of relevant information for incoming or prospective new students.

The Application Process

Applying to a graduate program can be quite a challenge. Obtaining support and guidance from a faculty member can help immensely in assembling an effective application. You will find that every graduate program has a specific application package with procedures and due dates that are unique to that school. Because you may wish to apply to more than one program, pay close attention to individual procedures and deadlines. Early application is always desirable, so this is another venture where delays or habits of procrastination will certainly work against you. Develop a simple system to ensure that the application materials remain organized. Begin to request transcripts of grades and required letters of recommendation as early as possible.

A particularly demanding aspect of the application process involves composing an autobiographical statement. Here again, consulting with trusted faculty members can prove extremely beneficial. In preparing your statement, keep in mind that the admissions committee members who review your application will view this autobiographical portion in multiple ways: as a sample of your writing ability, as an introduction to you as a person and a developing professional, and as a representation of your capacity to reflect on areas of personal growth and struggle. Many schools provide an outline or specific questions to which you must respond; a minority of programs provide only cursory directions and expect you to have some knowledge of what should be included.

The desired length for autobiographical statements varies from school to school, with an average range of 4 to 8 pages. Most commonly, you'll be asked to reflect on topics and questions such as the following:

Possible Items to Include in an Autobiographical Statement

Consider your decision to pursue a career in the helping professions. Describe how relationships with family, friends, and so on have influenced your decision to work with others.

What significant experiences have you had as a giver or receiver of help that have prompted you to enter this field of study?

In your life experience, how have you solved problems and struggled with important issues? In general, how has your value system been shaped by this process?

What personal qualities equip you for graduate education and for the profession you are seeking to pursue?

What social issues interest you, and why? What social problems exist in your home community that you would be able to address if you acquired an advanced degree?

Why are you applying to graduate school at this time?

What are your career goals and how did you decide on them?

Discuss your undergraduate academic experiences and include a description of your academic strengths and weaknesses. Explain any grade deficiencies and what you have done to improve them.

Graduate education often means a change in life style. What changes do you anticipate and how do you plan to prepare for them? What plans have you made to finance your education?

Discuss your current personal situation and any chronic health conditions or physical limitations as they relate to your full participation in graduate study and fieldwork.

As you can see, these questions are quite probing and require that you make careful decisions about what material is most appropriate to include. Many students struggle with the issue of how much personal information to include. It is easy to err on either side, submitting a statement that is too general to give the reader any real sense of who you are, or presenting too much personal detail, which may suggest inadequate boundaries. It may be helpful to invite several sources to review your material, thereby eliciting a broader range of feedback. Consultants also can help you refine your writing style, if necessary. You'll probably want to revise your statement several times until you feel reasonably comfortable with the finished product.

Selecting a Program

Your final selection will be the result of a process of elimination, determined in part by the offers you receive and in accordance with a variety of personal considerations. Decisions of this magnitude rarely are made without some degree of anxiety; however, taking steps to ensure that your decision-making process is careful and thorough will bolster confidence in your decision.

In making your final selection, we encourage you to think again about your reasons for wanting to attend graduate school and your overall aims and goals. Consider them in light of what each program has to offer. What do you feel passionately interested in learning? Does a particular program offer preparation in these areas? Master's or doctorate programs leading to the same degree title will vary in course content from school to school, depending on the predominant orientation, philosophy, or mission of the institution and the individual research interests of faculty members. We often encourage students to give greater attention to a program's particular coursework and fieldwork than to the title of the degree.

This is also the time to seek additional consultation from advisors or alumni of various institutions. Ask specific questions about the types of career positions that can be obtained with a given degree and the success rate of graduates who attended certain programs. Making sure the program is designed to fulfill your most important requirements is the best way to ensure a good fit and a successful outcome.

Obtaining Financial Aid

Most advanced degree programs offer financial aid packages once a student has accepted an offer to attend. Most students pay for graduate school with a combination of loans, grants, scholarships, and stipends. Investigate your options and speak with a financial aid advisor to gain a full understanding of what you must do to qualify. Pursue scholarship opportunities that may exist in your community and research corporations in your area that may sponsor students in advanced degree programs. Ask the university about loans that may be partially forgiven in return for post-degree teaching or employment in county social service and mental health sites. This process will require some investment of time on your part, but you may discover creative ways to finance your education.

Considering Professional Opportunities

Not everyone will desire to continue their education beyond the level of an undergraduate degree, or will be able to do so immediately following graduation. Economic factors may prompt a search for employment in a setting related to your educational training. Or, you may feel your chances for admission to a graduate program will be enhanced by a year or two of additional experience in the field, either in a paid or volunteer capacity. Some students appreciate the opportunity to take a break from academic demands and allow a period of time to pass before pursuing additional education. If you find yourself looking for employment as a new graduate, there are several key points to consider.

Do Your Research

Many universities have facilities that assist new graduates in their search for employment. These may include alumni advisors in the surrounding communities, a library of materials describing job search assignments, particular types of positions and their requirements, and regularly updated listings of available employment opportunities. It is wise to take an inclusive approach initially and remain open to learning about situations you may not have considered previously. Too much focus on a specific position or setting in the early stages of your search may limit your chances of finding suitable employment. Students who have completed one or more fieldwork assignments have established at least a few networking connections in their community. Although there may not be a timely opening at your previous fieldwork site, be sure to communicate your availability to former supervisors and colleagues who may have valuable leads for you to follow. The helping professions and human services agencies in a given community often form a matrix of interdependent resources; a connection to one source may open doors to many others.

Preparing an Effective Résumé

In your search for an acceptable position, a well-constructed résumé is usually essential for obtaining interview opportunities. Many guides are readily available to help you create a résumé that presents your professional experience and educational background in the best possible light. Computer software programs provide another means for devising this important document. (Former supervisors can provide valuable letters of reference and recommendation that can be accumulated in a file to be used as the need arises.)

Most new graduates are concerned about having a résumé that suffers from too little content. Because they are at the beginning of their professional careers, it is difficult to decide what elements of their employment history should be included or omitted. We encourage you to consult with others in your field and to seek the help of faculty advisors prior to graduating. Prepare a rough draft so that you will have something concrete to work from. You will find that you'll refine your résumé many times as you advance in your career. Experiment with wording and content, noting that the same position can be described in many ways with distinctly different effects.

At first, it can be quite useful to let a *specific* objective guide the development of your résumé. Once you have written a concise statement that describes the type of position you are looking for, it will become clear which elements of your education and experience you should emphasize. You can view a résumé as a document that *chronologically* details your suitability for the position you are pursuing, or you can arrange it *thematically,* according to different topics or areas of experience. Include all relevant information that supports the notion that you possess the skills and knowledge to successfully execute the duties of the position for which you would like to be considered. Omit or give only brief mention to previously held positions that don't relate to your current objective. The following list will help you prepare an impressive and persuasive résumé:

Tips for Preparing an Effective Résumé

- Describe employment experiences that emphasize specific skills that are relevant to the position targeted in your objective. Prospective employers are most interested in what you can *do.* It is up to you to describe your abilities honestly and precisely. Be sure to point out that which may not be obvious; even seemingly unrelated positions often provide the opportunity to develop and enhance skills that transfer nicely to other settings. For example, several years of experience in classroom teaching may provide a strong foundation for later work in a child and adolescent mental health setting.
- Include descriptions of your fieldwork experiences and any related volunteer or community service involvement. These may well form the body of your professional development at this time, so don't minimize their importance. It is not unusual for students to find that positions held as fieldwork students or in a volunteer capacity are identical to the paid staff positions they are now seeking.
- Make sure you have included relevant personal information, such as your current address and telephone number.

- Finally, assess the overall visual presentation of your résumé and invite the opinions of others. Spelling errors, poor grammar, and awkward or confusing phrasing will diminish your chances for consideration, especially if the position requires proficient writing skills, as many positions in the helping fields do. Ask yourself (and others) if the résumé is easy to scan. Does anything about the formatting distract the reader or detract from the information being presented? Is the type font you have selected befitting to a professional document? Is it readable? What is the quality of the paper you have chosen? Gray, white, or off-white stock looks the most professional.

See the box on page 121 for a guide to résumé preparation in the format of a typical résumé.

This chapter provides only a brief introduction to the many issues you are faced with at this point in your professional development. We hope we've given you a sense of how to approach these important decisions. Whether you decide to pursue further education or training, or seek employment in your chosen field, you will need to invest a great deal of time and energy to discover the path that is right for you. We wish you good luck in this quest!

Suggested Readings

Bernstein, G. S., & Halaszyn, J. A. (1989). *Human services? . . . that must be so rewarding.* Baltimore: Paul H. Brookes.

Bloch, D. P. (1994). *How to write a winning resume.* Lincolnwood, IL: VGM Career Horizons.

Collison, B. B., & Garfield, N. J. (1990). *Careers in counseling and human development.* Alexandria, VA: American Association for Counseling and Development.

Eyler, D. R. (1990). *Resumes that mean business.* New York: Random House.

Fortune, A. E., Feathers, C. E., Rook, S. R., Scrimenti, R. M., Smollen, P., Stemerman, B., & Tucker, E. L. (1989). Student satisfaction with field placements. In M. Raskin (Ed.), *Empirical studies in field instruction.* New York: Haworth.

Larsen, J. (1980). Competency-based and task-centered practicum instruction. *Journal of Education for Social Work, 16,* 87–94.

Raskin, M. (1989). Field placement decision: Art, science or guesswork? In M. Raskin (Ed.), *Empirical studies in field instruction.* New York: Haworth.

Russo, J. R. (1993). *Serving and surviving as a human-service worker.* Prospect Heights, IL: Waveland Press.

Guide to Résumé Preparation

OBJECTIVE

Write a clear, one-sentence description of the type of position you are seeking. The rest of your résumé should support this objective, demonstrating relevant qualifications for the position.

EDUCATION

Include your junior college and college degrees, or degree in progress with expected date of completion. You might also describe any special emphasis in coursework if it supports your objective (for example, major, minor, track within major).

EXPERIENCE

a. Separate paid from unpaid positions; professional experience from internships. Include community service or related volunteer positions.
b. Briefly describe, for each entry, what you DID. The emphasis here must be on skills required or acquired, responsibilities and assigned tasks, experience gained. Naming the position you held is necessary but does not tell a prospective employer what you did. This is your opportunity to paint a picture of yourself that effectively conveys your abilities. Choose your words carefully and state the information in a succinct manner.
c. Always present information in reverse chronological order, with your most recent experience listed first.
d. Remember that your résumé is the first representation of you. It speaks to the employer about your ability to identify and articulate your marketable skills. Ask yourself what your finished product says about you on first glance.

ADDITIONAL INFORMATION SOMETIMES INCLUDED

a. Awards/recognitions
b. Special skills/talents
   (for example, languages spoken, computer literacy, and so on)
c. Interests
d. Professional affiliations
   (student member of . . .)

REFERENCES

Rather than listing all of your references, simply state, "References are available upon request."

LENGTH

Try to be concise in your wording, using phrases rather than complete sentences. Strive to limit your résumé to one page (two only if absolutely necessary).

Index